CAMBRIDGE LIBRARY COLLECTION

Books of enduring scholarly value

Life Sciences

Until the nineteenth century, the various subjects now known as the life sciences were regarded either as arcane studies which had little impact on ordinary daily life, or as a genteel hobby for the leisured classes. The increasing academic rigour and systematisation brought to the study of botany, zoology and other disciplines, and their adoption in university curricula, are reflected in the books reissued in this series.

Studies in Bird Migration

Having trained as a civil engineer and surveyor, the ornithologist William Eagle Clarke (1853–1938) established himself in his field by preparing reports on bird migration for the British Association. Focusing on the species passing through the British Isles, Clarke spent many months in various lighthouses and on remote islands. He brought all his research together in this two-volume work, first published in 1912 and illustrated with maps and weather charts. In Volume 2, Clarke describes key examples of his investigations. Photographs of the sites he visited accompany the text. The locations range from the Flannan Isles, in the Outer Hebrides, to the island of Ushant, off the coast of Brittany. Clarke's expedition to the latter location ended abruptly when he and his colleague were mistaken for spies and forced to leave. Extensive coverage is also given to Fair Isle, between Shetland and Orkney.

Cambridge University Press has long been a pioneer in the reissuing of out-of-print titles from its own backlist, producing digital reprints of books that are still sought after by scholars and students but could not be reprinted economically using traditional technology. The Cambridge Library Collection extends this activity to a wider range of books which are still of importance to researchers and professionals, either for the source material they contain, or as landmarks in the history of their academic discipline.

Drawing from the world-renowned collections in the Cambridge University Library and other partner libraries, and guided by the advice of experts in each subject area, Cambridge University Press is using state-of-the-art scanning machines in its own Printing House to capture the content of each book selected for inclusion. The files are processed to give a consistently clear, crisp image, and the books finished to the high quality standard for which the Press is recognised around the world. The latest print-on-demand technology ensures that the books will remain available indefinitely, and that orders for single or multiple copies can quickly be supplied.

The Cambridge Library Collection brings back to life books of enduring scholarly value (including out-of-copyright works originally issued by other publishers) across a wide range of disciplines in the humanities and social sciences and in science and technology.

Studies in Bird Migration

VOLUME 2

WILLIAM EAGLE CLARKE

CAMBRIDGE
UNIVERSITY PRESS

University Printing House, Cambridge, CB2 8BS, United Kingdom

Published in the United States of America by Cambridge University Press, New York

Cambridge University Press is part of the University of Cambridge.
It furthers the University's mission by disseminating knowledge in the pursuit of
education, learning and research at the highest international levels of excellence.

www.cambridge.org
Information on this title: www.cambridge.org/9781108066983

This edition first published 1912
This digitally printed version 2014

ISBN 978-1-108-06698-3 Paperback

STUDIES IN BIRD MIGRATION

PLATE X.

THE KENTISH KNOCK LIGHTSHIP.

[*From a Drawing by Marian Eagle Clarke.*

STUDIES
IN
BIRD MIGRATION

BY

WILLIAM EAGLE CLARKE

Keeper of the Natural History Department, the Royal Scottish Museum

WITH MAPS, WEATHER CHARTS, AND OTHER ILLUSTRATIONS

VOLUME II.

LONDON

GURNEY AND JACKSON

EDINBURGH: OLIVER AND BOYD

1912

CONTENTS

LIST OF ILLUSTRATIONS

STUDIES IN BIRD-MIGRATION

CHAPTER XVIII

A MONTH ON BOARD THE KENTISH KNOCK LIGHTSHIP: A STUDY CHIEFLY DEVOTED TO THE EAST-TO-WEST AUTUMN MOVEMENTS ACROSS THE NORTH SEA.

AMONG the most interesting of the varied movements of birds observed in the British Isles and their vicinity are those remarkable intermigrations which take place in spring and autumn between the south-eastern coast of England and the opposite shores of the Continent, and which mainly come under notice at the numerous lightships stationed between the mouth of the Humber and the Straits of Dover.

If these important flights across the southern waters of the North Sea were not actually discovered through the investigations of the Migration Committee appointed by the British Association, it is assuredly owing to the labours of that body, and especially to those of my late and most intimate friend, Mr John Cordeaux, that attention was first seriously drawn to them.

During the preparation of the "Digest of the Observations on the Migrations of Birds made at Lighthouses and Lightships, 1880-1887," it became evident to me that much remained to be learned concerning these movements and the various species which participate in them, and I conceived the idea of undertaking some researches regarding them. To accomplish this, however, it was essential that I should spend some weeks on board one of the lightships—a course which demanded some consideration, since life on one of these floating observatories is inseparable from discomforts peculiarly its own. Encouraged, however, by the experience gained at the Eddystone lighthouse in the autumn of 1901, I decided to make the venture, and an application was forwarded on my behalf by the Royal Society to the Trinity House for permission to spend a month during the autumn of 1903 on board one of the Corporation's lightships in the North Sea. This privilege was graciously granted, and every facility was offered for visiting any vessel that might be selected.

The selection of a suitable station demanded careful consideration, and I finally decided upon the Kentish Knock lightship. This vessel appeared to me to lie at or near the centre of the migratory stream that I desired to investigate, and its remote situation, out of sight of land, promised to afford an excellent opportunity for witnessing the various movements, and the conditions under which they were performed, free from the influences which might prevail at stations nearer to our shores; lastly, the character of its light, a most important factor, seemed to be especially adapted for attracting the migrants which might pass in the night.

The geographical position of the lightship will be best realised by a reference to the map on page 4, which shows its situation in relation both to the English shores and those of the Continent. It is stationed in latitude 51° 38′ 50″ N., and in longitude 1° 39′ 55″ E., lying 21 miles north-east-by-north of Margate, and 21.5 miles south-east of the Naze, which are respectively the nearest points of land, while it is moored two miles east of the extensive sandbank from which it takes its name—a bank which is entirely submerged at all states of the tide. The following table affords some further information regarding its geographical relations :—

Direction from Kentish Knock.	Points struck on the English and Continental Coasts.	Miles from Kentish Knock.
North	A little south of Southwold on the Suffolk coast . . .	49
North-north-west	Mouth of River Deben on the Essex coast	42
North-west	The Naze on the Essex coast .	21.5
West	South-east coast of Essex near the mouth of the River Crouch	31
South-west	North coast of Kent near Reculver	28
South-south-west	East coast of Kent . . .	21
South	North coast of France a little east of Cape Griz Nez . .	53
South-south-east	Gravelines, on the north coast of France	48.5
South-east	Belgian coast near to the frontier of France	56
East	Mouth of the East Schelde, coast of Holland	88

The vessel is equipped with a white revolving light, throwing out three beams each of 12,000 candle-power, and making a complete revolution in three minutes. As it lies in the direct course of all the east-coast traffic passing north and south *via* the English Channel, and en route to and from the Thames, it is furnished with a siren of exceptional power, for use in times of fog or haze. This horn was a veritable *bête noire* to me during my early days on board the ship, though I afterwards, except during the night, came to regard its hideous and nerve-rending blasts with indifference.

Life in a lightship (apart from *mal de mer*, from which even the crew are not on all occasions immune) is undoubtedly one of considerable hardship and discomfort. It is the life of a seaman spent under the most trying conditions—namely, of one whose ship is ever the sport of the winds and waves. Off the Kentish Knock sands the set of the tide is so strong from north to south, and *vice versa*, that the lightship rides out gales from the east and west

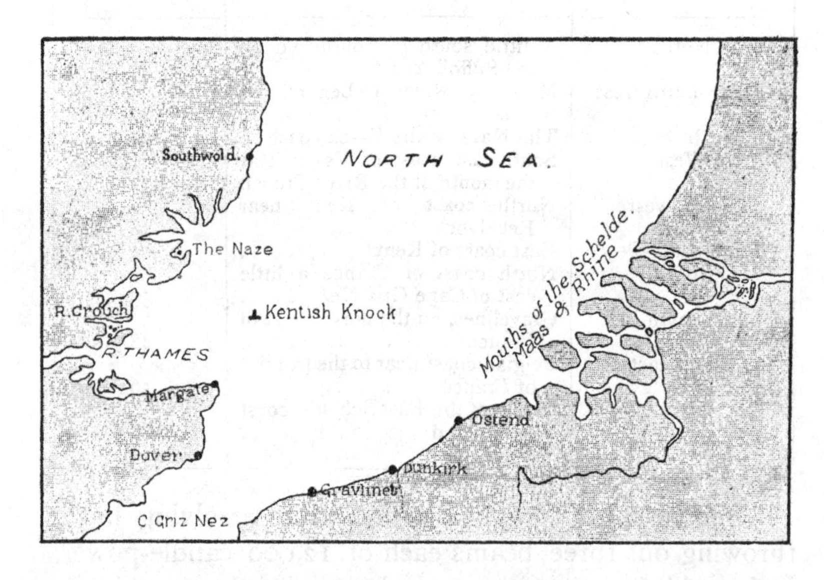

broadside to the waves. It was on these occasions that life on board became most trying. Then I was compelled to remain in my bunk, where I was sometimes so violently rocked in my cradle on the deep, that I found it decidedly difficult to avoid being ruthlessly tossed out of it. On these occasions, too, all the skylights were battened down, and artificial lights were burned below, and these, along with the galley close at hand, raised the temperature of the cabin to a degree that was almost intolerable. My only consolation on these occasions was the knowledge that I missed nothing, bird-migration being impossible. However, I enjoyed the best of health while on board, and the almost incessant watchfulness necessary for the successful prosecution of my work rendered my sojourn free from that extreme tediousness which would otherwise have been inseparable from residence in such a vessel.

I sailed from Blackwall in the Trinity tender "Vestal" on the morning of 15th September, and, after visiting the various lightships and "pile" lighthouses within the Thames estuary, and the outlying Galloper lightship, was placed on board the Kentish Knock lightship at noon on 17th September, and remained there until 18th October.

I found the bird-migration at the Kentish Knock of a very varied and complex nature, in which respect it is probably not surpassed by any other station on or off the British coasts. The ship lies about the centre of a broad junction where many lines of flight cross each other—a result of the close approximation of the British and Continental land masses just to the south of it. Here, in addition to (1) the extensive movements (I speak of those of the autumn, the spring migrations being in an opposite direction) of immigrants from east to west and south-south-west and north-west, there are (2) movements of a similar nature from south-east to north-west, and (3) of birds of passage along the first-named route; while (4) emigrants, including many birds of passage, pass from north to south-south-west, and (5) from north-west to south-east. There appeared to be no Continental migration whatever from points north of east. It will thus be realised that much "cross-migration" takes place, and this, too, strange to say, is sometimes performed by identical species on the same day and even at the same hour. Another feature of importance is that at the Kentish Knock and neighbouring lightships the day movements equal, if they do not surpass in magnitude, those observed during the night, whereas at other stations around our coasts the nocturnal movements vastly exceed in extent those occurring during the day.

As at the Eddystone in 1901, I found it extremely difficult to detect small birds travelling singly or in twos and threes during the daytime. The great majority—nearly all of them, in fact—flew close to the water, and the surface, almost always in motion, forms a most unsatisfactory background against which to pick up migrants, unless they are passing in large flocks. The rougher the sea the more difficult is the task of observation, and the higher the wind the more closely do the birds hug the surface of the sea ; thus, except during a dead calm, many migrants escape notice, in spite of the utmost watchfulness on the part of the observer.

In connection with the movements witnessed at such isolated stations, it must be remembered that these observatories are mere specks in the open sea, and the marvel is that one sees so much, especially during the daytime. At night it is somewhat different, for then, under certain atmospheric conditions, numbers of birds are attracted to the lantern—many of them, no doubt, being allured from afar.

In the preparation of these results I have grouped the observations under the particular set of movements or problems to which they relate, reserving certain information to be dealt with under the various species.

The first migratory movements to come under notice

were those of birds seeking winter homes far to the south of the British Isles. I was much gratified to find that, although such an outlying station, the lightship lay in the course of the southerly passage of numerous summer birds departing from the more northern counties of Great Britain, as well as from Northern and Central Europe. Many migrants from the north when skirting our shores find themselves far to the eastward on reaching the coast of Suffolk, and on leaving that county proceed over sea towards the east coast of Kent, a course which carries them near the lightship, where not a few of them were observed proceeding to the south-south-west, as did others which had travelled westwards from the opposite shore of the North Sea.

Numbers of such emigrants passed between 18th and 29th September—a genial spell of weather, with much sunshine and light breezes, following a particularly cold and stormy period for the time of the year. Between these dates many Wheatears, Redstarts, Skylarks, Pied Flycatchers, and Tree-Pipits flitted by; and fewer Meadow Pipits, Starlings, Goldcrests, Pied Wagtails, Yellow Wagtails, and Swallows, all *singly* except the Skylarks and Swallows, which passed in small parties. These migrants arrived singly and not unfrequently followed each other in quick succession, but there were usually greater or lesser intervals between their appearances. Not a few alighted on the ship, where some of them being both tired and hungry, spent a considerable time resting or busying themselves in an active search for insects, of which we had numbers on board at the time.[1]

[1] For an account of some of the insects observed on board the Kentish Knock lightship, see the *Entomologist's Monthly Magazine* for December 1903, p. 289.

Others remained for a few moments only, and then took their departure. In all cases these birds on leaving the ship winged their way towards the coast of Kent.[1]

No doubt many birds of the species named and others passed without coming under notice, for a very slight deviation to the west, or east, would carry them beyond the range of observation. Among the rarer species observed were an Icterine Warbler and a Blue-headed Wagtail. The latter, in common with many other species, most probably arrived from the east, though it was in most cases impossible to tell from what quarter these birds came, for they appeared, as it were, mysteriously, not being observed until they perched on the rail or rigging.

A Phalarope, probably *Phalaropus hyperboreus*, was observed on the water some little distance from the ship on 13th September. It was one of the very few Limicoline birds that came under observation, and was the only one detected during the daytime.

These movements of summer birds on their way southwards were prolonged beyond the limits of September. Thus Wheatears and Chiffchaffs were observed on 3rd October; Starlings, Chaffinches, and Swallows passed on the 13th; Sand-Martins on the 15th, and Swallows again on the 16th. Here, too, may be mentioned the Rock-Pipits observed on 23rd September, and on 8th and 12th October.

[1] The 19th of September was a great day for migrants (probably most of them immigrants from North-Western Europe) on the coasts of Lincolnshire and Norfolk, where, as I am informed by Mr Gurney, Redstarts, Pied Flycatchers, Redbreasts, Goldcrests, Ring-Ouzels, Lesser Whitethroats, Bluethroats, Blackcaps, and Grasshopper Warblers occurred. On the same day Redstarts, Pied Flycatchers, Wheatears, Willow-Warblers, and Tree-Pipits were passing south-south-west during the afternoon at the Kentish Knock.

The day movements were chiefly observed during the forenoon, but on some occasions were continued until sunset; while the hour and duration of the nocturnal visits entirely depended upon the advent and prevalence of weather conditions suitable for bringing night migration under observation.

Continuing the observations relating to emigration, I have next to notice a series of movements from the north-west towards the south-east—that is to say, from the Essex coast, at or about the Naze, to the Belgian coast near the French frontier—a line of migration which seems to have been overlooked, but one to which I have recently drawn attention when treating of the spring migrations, in the reverse direction, of the Rook and Starling. The emigrations observed along this route, though marked, were confined to a few species, but it formed the main line across this part of the North Sea, by which the House-Martin, Meadow-Pipit, and Pied Wagtail sought the south, and was also a minor route for Wheatears, Starlings, and Skylarks. Considerable numbers of the three first-named species traversed it on 7th, 9th, and 14th October, days on which there was little or no migration along other lines of flight.

These movements to far southerly winter quarters were by no means confined to the daytime, but were in progress at night, when the weather was favourable for migration and the atmospheric conditions for their observation. All the species already mentioned (excepting the Wagtails), with the addition of Common Whitethroats, Spotted Flycatchers, Thrushes, and Blackbirds, were observed around the lantern, sometimes for several hours and in great numbers.

We now approach the East-to-West flights, which are certainly the most interesting of the migrations observed, and are also the main diurnal oversea movements regularly witnessed on the British shores.

The lightship, I found, occupied a central position amid this great feathered stream, since the vast majority (90 per cent. or more) of these migrants were moving direct from east to west, while others flew to the south-south-west and north-west. At stations off the Norfolk coast their chief line of flight is to the north-west, off the Lincolnshire coast to the north-north-west, while off the east coast of Kent it is to both west and south-west.[1] On certain days, when general movements are in progress, these flights have often been known to cover the entire coastline between the Humber and the Straits of Dover.

On arriving on our shores many of these immigrants proceed inland and settle down for the winter, while others traverse the south coast and cross St George's Channel to winter in Ireland; others, again, cross the English Channel en route for Southern Europe.

Owing, perhaps, to the unexpected spell of summer weather that characterised the latter half of September, and to the high and uniform temperatures which prevailed then and during the first week of October, the east-to-west movements of the autumn of 1903 were very slightly in evidence in the earlier days of my residence on broad the lightship. During the period indicated only a few Skylarks, Tree-Sparrows, Swallows, Meadow-Pipits, and single Starlings were seen. The night movements of waders, however, appeared to be of

[1] For general information on these lines of flight, see Vol. I., pp. 83-87.

more importance; but, alas! only the notes of Ring-Plovers and Lapwings could be identified among the many voices that reached me as the migrants sped westwards under the cover of darkness.

The first extensive movement immediately followed a decided fall in temperature on the Continent. This commenced at 9 A.M. on 8th October, and from that hour until 2 P.M. flock after flock of Skylarks and Chaffinches and small parties of Tree-Sparrows and Meadow-Pipits followed each other in rapid succession. Starlings, which had hitherto only been noted singly, also passed in small troops. It was an important morning for east-to-west migration, and not only did hundreds of birds pass quite close to the ship, but far greater numbers, in fact many thousands, were observed pursuing a like course at distances too great to render their identification certain, especially amid the dull weather and heavy rain which prevailed, and from the fact that all were flying close to the surface of the sea.

On 10th October there was another considerable fall in temperature, and our thermometer registered 10° lower than on any previous occasion since my residence in the lightship. This was followed on the 11th by the greatest diurnal movement of birds that I have ever witnessed. It set in at 8 A.M. with a marked passage of Starlings, Skylarks, and Tree-Sparrows. By midday it had assumed the nature of a "rush," which was maintained without a break until 4 P.M. It was a remarkable movement in many ways. Skylarks, Starlings, Chaffinches, and Tree-Sparrows not only passed westwards in continuous flocks, but many of these companies consisted of hundreds of individuals. So numerous were the Starlings composing some of these bands that when first observed in the distance they resembled dark clouds, and formed a conspicuous contrast to the leaden, white-crested billows. The elements contributed to the singularity of the scene. The weather, which had been fine up to 9 A.M., rapidly changed, and by noon it had become, in nautical parlance, a "dirty day"—a character which it maintained to the end. The rain, which fell steadily at first, became a downpour, and finally torrential. Indeed, so rain-laden did the atmosphere become, that it was necessary to sound the fog-horn, whose hideous yells added a weird accompaniment quite in harmony with a scene which, apart from its intense interest to a naturalist, was dismal and depressing in the extreme. The wind, too, had been gradually rising, and by 3 P.M. had increased to a "strong breeze" with a velocity of 34 miles an hour. There were squalls at intervals, which lashed the rain against my face with such violence as to cause the skin to tingle for a considerable time. How the migrants braved such a passage was truly surprising. How they escaped becoming waterlogged in such a deluge of wind-driven rain was a mystery. Yet on they sped, hour after hour, never deviating for a moment from their course, and hugging the very surface of the waves, as if to avoid as much as possible the effects of the high beam wind. It was surely migration under the maximum of discomfort and hardship, indeed under conditions that approached the very verge of disaster for the voyagers.

It is probable that the birds would not have quitted the Continent had these later conditions prevailed at the hour of their departure. That they did not do so is made clear by a reference to the "Daily Weather Chart" issued by the Meteorological Office, and also by the observations registered at the lightship. The fact is that the weather changed rapidly under a falling barometer and a southerly wind; and thus, although the migrants set out under favourable conditions for the passage, they were overtaken while en route by the changes which became more and more unfavourable as they neared the English coast, and approached more nearly the storm-centre which lay off our own western shores. Thus were they trapped, and had to make the best they could of a bad passage.

There were also westward flights of considerable magnitude on the part of the same species on the following day, 12th October, and again on the 15th. These were performed under conditions which were not unfavourable to the migrants. On the latter date some remarkable cross-migrations were observed on the part of Skylarks and Chaffinches, passing flocks of which were coming from both the north and east, sometimes simultaneously, during the morning.

On 17th October, Skylarks and Starlings were passing west at intervals during the day, in spite of a somewhat high northerly wind. Late in the afternoon the first Rooks and Jackdaws appeared in small numbers, as they did also, as I afterwards ascertained, at the Galloper lightship, a vessel moored over thirteen miles to the east-north-east of us. The advent of these birds was of great interest to me. I had been expecting them for some time, for they were overdue. Several individuals of each species appeared at 4.30 P.M. and alighted on the ship, but did not remain long ere they departed westward. At 6.30 P.M., in continuance, no doubt, of the movement referred to, several Rooks and Jackdaws appeared at the lantern and flew around for some hours, indeed until daybreak the following morning, and one adult Rook and two Jackdaws were captured. A great number of Starlings, Skylarks, Chaffinches, Mistle-Thrushes, Song-Thrushes, Goldcrests, Meadow-Pipits, Wagtails, and other undetermined passerines were present, and doubtless also came from the east.

The 18th added some interesting experiences. The "relief" at the Kentish Knock lightship was effected by the Trinity yacht "Irene" at 9 A.M. A great westerly movement was in full swing at the time, under weather conditions which were eminently favourable to the migrants; these comprised the usual species—namely, Starlings, Skylarks, Chaffinches, and Tree-Sparrows. At 10.30 A.M., we bade adieu to the lightship and steamed south-west towards the Kentish coast, on nearing which our course was changed, and we proceeded west in the direction of the Thames estuary, and finally to the mouth of the river, where, at 3 P.M., I was put ashore at Southend.

During the entire passage of four and a half hours—the distance travelled being close upon fifty miles—we were at first crossing the course of, and afterwards running parallel to, the flight of continuous flocks of Starlings and Skylarks, and of fewer Chaffinches and Tree-Sparrows, all proceeding westwards, and all flying just above the surface of the calmest of seas and in the finest of weather.

These flocks, especially those of the two first-named species, were never absent from view, and we must have encountered tens of thousands of these birds during the passage. It was a revelation even to one—shall I admit it?—painfully familiar with the voluminous records of such movements chronicled in the migration schedules; but it is one thing to study in cold blood, as it were, masses of statistics, and quite another to witness these bird-streams actually flowing unceasingly before one, hour after hour. It was the marvellous continuity and apparently inexhaustible nature of these movements that were a revelation to me, both on this and other occasions.

The flocks ceased to be so numerous as we approached the mouth of the Thames proper, but groups of Starlings and Skylarks were still moving westwards when I left the "Irene" at Southend at 4 P.M.

A Mistle-Thrush, observed flying somewhat high and to the west soon after we left the Kentish Knock, was the only Thrush that came under my notice during the daytime.

In addition to the species named as participating in the great diurnal movements I have endeavoured to describe, Swallows and Martins in considerable, and Wheatears in lesser, numbers were also observed moving westwards independently. The fact that these species proceed along this route in autumn is the clearest possible evidence, if such were necessary, that it is also a true passage fly-line for migrants proceeding from the Continent *via* the English shores to their winter quarters south of the British Isles. These birds of passage, after arriving in south-eastern England, sooner or later take their departure from our southern coast, en route for Africa in some cases, and South-Western Europe in others. Occasionally during the daytime small numbers of the summer birds just mentioned, remarkable to relate, were proceeding from south-east to north-west: proceeding north ere they sought the south! but this was exceptional, and the movements were never important.

It will have been noticed that the species recorded as participating in these great east-to-west movements are comparatively few in number—at any rate, those observed during the daytime. At the Eddystone, likewise, a few species only cross the Channel by day, but a great variety at night. May not the same be the case at the Kentish Knock? I am decidedly of opinion that this is so. I am inclined to think that we have here the true explanation of what takes place. The phenomena of migration as witnessed at this station are, however, exceedingly complicated, as has already been stated, and it was found impossible to determine from actual observation whence came the birds that were so abundantly noticed during the hours of darkness. The occurrence of such characteristic east-to-west migrants as Rooks and Jackdaws at night, also favours this opinion. I shall return to this subject anon when discussing the night movements.

The reason why the movements of these species are performed so largely during the daytime, instead of entirely at night, as is the case of most over-sea migrations, may be the comparative shortness of the passage (about 180 miles at most), if direct east to west in the latitude of this lightship. The few hours necessary for its accomplishment would not interfere unduly with the time that must be devoted to the daily search for food—a most important consideration for all migrating birds.

It is a fact worthy of mention that each of the flocks,

great and small, that came under observation during these great movements, was composed of a single species. I never saw a flock or party consisting of mixed species—each kind kept strictly to itself, even when passing simultaneously with others and moving in the same direction.

An interesting problem in connection with these east-to-west movements is: Whence came these vast hosts of autumn migrants—Continental emigrants? I was somewhat uncertain when I prepared the "Digest" in 1896. Now I am decidedly of opinion that they are birds of Central (Western) European origin. I have little doubt of this from their lines of flight, from their species, and other considerations. These migrants, I believe, quit the Dutch coast at the mouths of the Maas, Rhine, and Schelde, which they have reached mainly by pouring down the courses of these great rivers from inland districts, some of which may lie far away in the interior of Europe. This would account for the vastness of their numbers.

The extraordinary persistency with which these hosts follow definite lines of flight during their passage across the North Sea is very remarkable and also bears out my view. Thus at lightships lying only a few miles off the coast, and well within sight of land, the birds are *not* recorded as making for the nearest points of the land, but as persistently following particular lines of flight. It is quite reasonable to suppose that the same definite course has been maintained during the entire journey; and if we trace such lines back to the shores of the Continent, we shall find, whether the observation be made off the coast of Lincolnshire or Kent, that they have their origin on that section of the coast of Holland which I have indicated.

I was much struck with the small number of essentially marine birds that came under my notice at this pelagic station. The only Gulls that were fairly numerous were the Lesser Black-backed and the Kittiwake. The Great Black-backed Gull was occasionally observed, but only a single Herring-Gull came under notice. Skuas, chiefly Arctic, and a few Pomatorhines, were frequently in attendance on the Gulls. I saw one Tern—an immature example of the common species. There were no Shearwaters, but I saw a single Fulmar. Gannets, all adults, were not unfrequent as passing visitors. Common Scoters, Guillemots, and Razorbills were numerous along the edges of the sands, and were chiefly in evidence during the prevalence of strong westerly winds, when they sought our side of the banks —the easterly—for shelter and food. I also saw a few Red-throated Divers.

One of the crew of the lightship, who has been on board for over a year, and who, being a bird-fancier, knows all the small cage-birds well, informed me that he had never seen a Goldfinch, Linnet, or Redbreast on or about the vessel since he had been stationed there. He had seen two Greenfinches, which had arrived together during the spring.

The night movements were of a varied and interesting character, occasionally of considerable magnitude, and problematical in their nature.

I may say at once that, so far as direct observation was concerned, it was on all occasions quite impossible to tell from what quarter the birds approached the ship after darkness had set in. This I much regretted, for I was particularly wishful to ascertain whether the east-to-west movements were performed during the night as well as the day. I did ascertain beyond a doubt that Rooks, Jackdaws, Lapwings, Ring-Plovers, and a number of wading birds did move in this direction during the night-time, and I think that there is strong presumptive evidence that the great movement on the night of the 17th to 18th October was chiefly if not wholly from east to west, and I believe that birds from the east were present on several other occasions.[1]

The nocturnal passage southwards during the latter part of September has already been alluded to, and but little more remains to be said concerning it. On the night of 18th September and during the earliest hours of the 19th, Redstarts, Pied Flycatchers, Thrushes, and an adult male Kestrel were at and around the lantern, along with other species unidentified; and several Common Sandpipers were heard passing, but did not show themselves. This was probably the commencement of a movement southward which was in full swing all the following day.

Soon after midnight on 20th September a large party of Skylarks appeared, accompanied by other small passerines. A considerable number struck the lantern and fell into the sea, the wind being moderately strong and the ship riding with her beam to it.

At 8.45 P.M. on 22nd September a number of Wheat-

[1] It is a very significant fact (one which favours the opinion that many, perhaps most, of these night migrations were from the east westwards), that although I witnessed movements as late in the season as 18th October, yet I never saw a single essentially northern species, such as the Redwing, during the whole of my residence in this lightship. At the Galloper lightship, east of the Kentish Knock, on 22nd October 1887, Rooks, Starlings, and Larks were at the lantern all night, clearly proving that these east-to-west passages are performed during the hours of darkness.

ears arrived and continued to fly around while the sky remained overcast.

On 25th September, between 1 and 4 A.M., during light rain, many Wheatears, Redstarts, Pied Flycatchers, Whitethroats, Willow-Warblers, Tree-Pipits, Skylarks, and a Richard's Pipit (examples of all of which were killed at the lantern), doubtless with other species, were flying round the vessel, and great numbers struck the glass and were lost in the sea. The presence of Richard's Pipit, a Central European summer bird, suggests that this movement was, at least in part, from east to west.

On 29th September, between 2 and 5 A.M., Blackbirds (those killed being immature males), Redstarts, Pied Flycatchers, Wheatears, Goldcrests, and Skylarks were present in great numbers, and hundreds struck the lantern and fell overboard.

On 3rd October, from 1.30 to 4 A.M., Goldcrests, Mistle-Thrushes, Song-Thrushes, Blackbirds, Chiffchaffs, Meadow-Pipits, Wheatears, and others were flying round. The rays of light were not very brilliant, however, and comparatively few perished at the lantern.

The fortnight that followed was devoid of night movements. There were days on which much east-to-west migration was witnessed, but the nights were birdless, so far as observation was concerned, for the weather conditions were not such as to render the lantern attractive to passing migrants.

From 6.30 P.M. on the night of 17th October to 5 A.M. on the 18th, Starlings, Larks, Chaffinches, Jackdaws (a few), Rooks (a few), Mistle-Thrushes, Song-Thrushes, Wagtails, Goldcrests, Meadow-Pipits, and probably many other species were careering around the ship, and examples of those named were either killed or captured at the lantern. This was the most important of the night movements witnessed, for some thousands of birds struck the lantern and fell overboard during the ten and a half hours that it was in progress. In this movement, I think, we have most unmistakable evidence in favour of its being, in part at least, an east-to-west migration. It commenced as soon as it was dark, and some of the species participating in it, notably the Rooks, Jackdaws, Chaffinches, Skylarks, and Starlings, had been observed moving in that direction, as long as it was possible to see them in the gloaming—that is to say, down to within an hour, or a little more, of the first appearance of the birds at the lantern. Thus it must be regarded as a continuation westward of the flights witnessed during the day. The presence, too, of the Rook and the Jackdaw, and the entire absence of any essentially northern species, must be considered as favourable to this view.

These night movements were very interesting to witness, and were novel to me, since they were seen from an entirely new standpoint—namely, from below. Of these new experiences, perhaps the one which impressed me most was that, from the deck of a lightship, one realised more fully the terrible loss of life that is involved by these nights at the lantern. Here one saw birds actually falling thickly around, and even heard them dropping on to the surface of the water. Such scenes often lasted for hours—ten and a half hours on the 17th to 18th October—and the sacrifice of life on this and other occasions was simply appalling. Some of the victims, indeed the majority, were only stunned or slightly injured, and thus met with a miserable death at sea. Few fell on board, unless the night was still, and then chiefly those which struck the lantern with considerable force, and fell like stones below.

Seen from the deck, the three beams from the lantern appeared to be thrown towards the surface of the surrounding waters at an angle of 45°. The birds—brilliant glistening objects—seemed to ascend, as it were, these streams of light by a series of short jerky flaps performed by wings which appeared to be only half spread for flight. Some of them paused when within a short distance of the lantern, remaining almost stationary, as if to sun themselves in the radiance of the slowly passing beam. Others were bolder and approached the light more closely, but ere they reached it spread their tails like fans, in order to check, at the last moment, their perilous onward course, and then sheered off, returning in a moment or two to repeat the performance. This spreading of the tail was a pleasing trait, especially in the Wheatear, whose black-and-white rectrices formed a very pretty fan. Others, again, approached the light gently, and either fluttered against the glass, or, as was particularly the case with the Starling, perched on the iron frame-work of the lantern-windows and seemed to revel in the light. In this respect the Starling differed from the rest, and when one brilliant beam had passed, the bird craned its neck and appeared to gaze longingly towards the next, which was slowly approaching. Indeed, the actions of the Starling in particular showed the birds under the spell of some overpowering fascination. A number of the visitors made their *début* with a wild dash for the light, and these, if they struck the glass direct, were killed outright; while if the contact were made obliquely, they glanced off stunned, and, slightly injured, descended with a curious zigzag flight which sometimes carried them some little distance ere they were lost amid the waves. The Rook cut the sorriest figure of all the migrants seeking the light. He too tried to obtain foothold on the frame of the lantern, whereon to sit amid the blaze of light, but failed, and flapped and struggled against the windows in a singularly clumsy fashion. Finally, to complete the scene, there was the singular effect produced by its central feature—namely, the great lantern, which, placed high up on the mast, swung slowly to and fro amid the glittering hosts that danced attendance upon its mystic charms.

On occasions when the rays were not particularly conspicuous the migrants flew aimlessly around, passing from ray to ray, sometimes for many hours. It is extraordinary how long some birds will fly round a light without resting. As a good example may be mentioned the case of a Kestrel which appeared at 8 P.M. on 18th September, and careered around without a break or rest of any kind until 1.30 A.M. on the 19th. This bird often came close up to the light, but checked itself by spreading its tail; and it also frequently flew to windward, and then dashed back over the lantern at a tremendous pace. It paid no attention to the few birds which were sometimes present during its prolonged visit.

When the wind was somewhat high, the birds resorted almost entirely to the lee side of the ship, and approached the lantern head to wind.

Although some of the night movements witnessed were of considerable magnitude and remarkably prolonged, yet the migrants, on the whole, were singularly silent—indeed, disappointingly so, for thus a useful aid to identification was denied me. Nor could I walk around the lantern and inspect them, as they fluttered against the glass. A novel method for the capture of specimens for determination was, however, resorted to with considerable success: a sailor was stationed on the sloping roof of the lantern, where, armed with an angler's landing-net, he captured the birds, like so many moths, as they streamed up the beams of light towards him. In this way many birds, ranging from a Goldcrest to a Rook, were secured for the purpose of identification.

The weather conditions, under which the rays from the lantern became conspicuous and attractive, were identical with those I had noted at the Eddystone (see Vol. I., p. 284)—namely, the existence in the atmosphere of moisture not necessarily in the form of rain or haze, but actually present, thoughnot visible, on dark starless nights. In order to put my views on this subject to a scientific test, I took with me to the lightship a hygrometer, with the object of ascertaining the actual percentage of humidity in the air on such occasions. I had not many chances of using the instrument, as either rain or haze was, in most instances, present, but on several occasions when it was not so manifested, I found the hygrometer indicated a very large percentage of moisture, in two instances reaching as high a figure as 86—a more pronounced result than I had anticipated.

There was hardly a single occasion during my visit on which the rays were conspicuous and the birds absent; on the other hand, there was not a single instance of migrants visiting the light when the night was bright and starlit or the moon was visible.

The birds which appear at the lantern are, by some authorities, considered to be birds that have lost their way, and hence make for the light in default of any other directive impulse. After my experiences at the Eddystone and the Kentish Knock, I am convinced that this is not the true explanation. I believe that the migrants are actually decoyed from or arrested on their course by the influence of the light itself. At the Eddystone, the emigrants which I saw in such numbers had practically only just left the land behind them, and had not had time to get lost when they appeared at the lantern. Another important fact in support of my contention is that the birds never appear at the light-stations at night except when the rays are remarkable for their luminosity; and in this connection it is important to bear in mind that this brilliancy does not depend upon such a thickening of the atmosphere as would cause inconvenience to the birds during their passage, for I have seen them in great abundance at the lanterns when I could make out neighbouring lights that were ten miles or more distant. Another significant fact is that they do not seek stations having red or green lights. Such lanterns, I am informed by the keepers, are seldom if ever visited under any conditions, for, owing to the subdued nature of their lights, the rays never become sufficiently conspicuous to prove attractive. When the Galloper lightship had white lights, great numbers of birds were allured to its lanterns, but now that the light is red, bird-visitors are almost unknown. If the birds were lost, why should they seek a white light and avoid one that is red or green? That the migrants may and do become confused, and for a time, perhaps, lost after the excitement and fatigue occasioned by their attendance upon the lantern, I can well imagine.

There is another very remarkable fact concerning these visitors to the lights to which I have never seen any allusion made—namely, that *the vast majority of them are passerines!* I have seen tens of thousands of migrants around the lanterns of the Eddystone and Kentish Knock stations, and all were passerines except two—namely, a Storm-Petrel and a Kestrel. And yet I have heard waders and other birds passing during these stirring nights at the lantern, though beyond giving tongue they passed by unconcerned and invisible. How are we to account for this? Why should the Passeres be allured to the light and not the Limicolæ? Can it be because the former—the most specialised of birds—are rendered, by reason of their higher organisation, more susceptible to the mysterious influence of the light? I merely throw out this suggestion as a possible explanation.

As to the meteorological aspects of the migration phenomena witnessed at the Kentish Knock, not much remains to be said, for frequent allusions have already been made to them when treating of particular movements. In dealing with this section of the subject, I have consulted a set of the "Daily Weather Reports," issued by the Meteorological Office, wherein are shown the conditions prevailing over the whole of the western half of Europe. The conditions most favourable for passages across the southern waters of the North Sea prevail when the central area of an anticyclone lies to the east of our islands, when, as already explained (see Vol. I., p. 173), the winds would range from south to east.

The main weather-influences for investigation were naturally those associated with the east-to-west movements. These movements, I found, did not begin in earnest until a decided fall in temperature took place in Western Central Europe, and this important factor was the precursor of each of the pronounced movements observed. Before such incentives to migration were experienced, unusually high temperatures had prevailed, and this was undoubtedly the reason why the movements prior to 8th October had been of such a straggling and feeble nature. These falls in temperature were not on all occasions experienced on our shores, and this again demonstrates the necessity for consulting the meteorological data at the place where such movements have their source.

The strongest wind prevailing when migration was in progress was on 13th October, when, with a westerly moderate gale blowing with a velocity of from 34 to 40 miles an hour, Swallows were proceeding in numbers to the south-south-west, and some House-Martins to the north-west—no other species being on the move. Under like conditions on the 9th, Martins were the only migrants observed, and were moving from north-west to south-east.

The weather conditions under which the other movements were witnessed do not call for any special remarks, for my experiences were similar to those at the Eddystone, and supported the views already expressed in that "Study" on the bearing of meteorology on bird-migration where the sea-passage is a short one.

As at the Eddystone, whenever a number of individuals of a species were obtained during any movement, they showed a considerable range of variation in their wing measurements, bearing out fully what I have previously said on this subject (see Vol. I., p. 307). As showing how much individuality may enter into these measurements, it is of interest to note that in six Wheatears, all females in identical plumage, killed or captured on the early morning of 25th September, the range of wing varied from 3.62 to 3.88 inches. It may also be noted that of twenty-two Skylarks obtained on 29th September, the wing range was only from 3.78 to 4.35 for young and old males and females, the average being 4.07 inches; while of ten examples captured on 17th to 18th October it varied from 3.93 to 4.70 inches, the average being 4.3 inches.

The height at which birds fly when migrating is a subject on which much has been written, and the fact that they have been observed proceeding at considerable elevations has been advanced as explaining the mystery of their being able to find their way, especially when a considerable extent of sea has to be crossed. That some birds do fly at great heights, and that under certain conditions (probably during fine weather) it may be an advantage to them to do so, I will not for a moment deny, but I am convinced that it is not a necessity as a means of finding their way.

The birds observed crossing from east to west in the latitude of the Kentish Knock would have a flight of at least 120 miles to perform between the Continental and the English coasts. When observed at that lightship, they had over one-fourth of their flight still before them, so that it was an excellent station for studying this and the various other conditions under which the journey was performed. During all these movements, great and small, the migrants of every species flew close to the surface of the water under all conditions of weather. On certain occasions, notably on 11th October, the state of the atmosphere was such that it must have been quite impossible for them to see more than one, or at most two, hundred yards ahead; and yet under these conditions, when it might possibly have been an advantage to fly high, they sped onwards, just skimming the crests of the waves, and never departing from a true east-to-west course. On fine clear days, with a light wind, these flights were performed in a precisely similar manner. Such facts as these, as well as many others, afford, I think, conclusive evidence that birds are endowed with a sense of direction which, even under exceptional circumstances, seldom fails them.

The speed at which birds fly while actually on migration is another moot point on which I was able to obtain some information at this remote station in the North Sea. Speaking generally, the migrants pursued their way at the steady rate characteristic of their respective species. There was no hurry, but at the same time there was a business-like manner about them which was in keeping with the important event on hand. Certain species habitually fly faster than others: thus the flight of the Meadow-Pipit was slower than that of the other species observed; that of the Skylarks, Chaffinches, Wagtails, and others was decidedly faster; while that of the Starlings, Martins, and Swallows was the speediest of all.

I had an excellent opportunity for roughly gauging the speed of both Skylarks and Starlings on 18th October. The "Irene" ran for some hours in the same direction as the flight of these species. Flock after flock flitted alongside of the ship, and, at my request, the captain ascertained from the engine-room the precise rate at which we were steaming, the result being a speed of exactly eleven knots (12.6 miles) an hour. At this speed the Skylarks passed us with the greatest ease, and, as near as it was possible to estimate, were proceeding as fast again as the ship, or at a rate of about 25 miles an hour, but certainly not more. It was more difficult to estimate the speed of the Starlings, but they were probably travelling at least half as fast again as the Larks, and therefore at not less than from 35 to 40 miles an hour.

The following list affords, in an epitomised form, and for each of the species observed, the information relating to their various movements and the dates on which they were performed :—

TURDUS VISCIVORUS, *Mistle - Thrush.*—With one exception, a night visitor, observed in some numbers at the lantern and around the ship on the early morning of 3rd October, and again on the night of the 17th and in the early hours of the 18th, examples being captured on both occasions. One was flying high to the west at 11 A.M. on the 18th.

TURDUS MUSICUS, *Song-Thrush.*—A night visitor only. Numbers were flying around the ship between 2 and 4 A.M. on 18th September and 3rd October, some being killed on the latter date. Many were again present during the great night movement of 17th-18th October, when several were killed at the lantern.

TURDUS MERULA, *Blackbird.*—Was observed between 2 and 4 A.M. on 27th September and 3rd October, along with other species. On the former date two were killed against the lantern, both immature males.

SAXICOLA ŒNANTHE, *Wheatear.*—This bird was observed migrating on ten days. It was noticed singly, but numerously, during the daytime, on 19th, 20th, 22nd, and 27th September, and 1st October; and at the lantern on 22nd, 25th, and 29th September, and on 2nd and 3rd October. On a few occasions single birds were seen flying to the south-east, evidently en route for the coast of Belgium or France; and some, likewise singly, to the north-west. When attracted to the light, however, the bird was present in numbers.

RUTICILLA PHŒNICURUS, *Redstart.*—Passed to the south-west on seven days between 18th and 26th September, and was also numerous after dark. During the daytime it was observed to flit by, singly but continuously; and at night several were at the lantern simultaneously, though they may have appeared singly. On 25th September, a beautiful pinkish-buff variety, with paler under surface and almost white wings, but with normally coloured lower back and tail, was killed at the lantern at 2 A.M., and proved to be a young male.

SYLVIA SYLVIA, *Whitethroat.*—Participated in the great migratory movement witnessed on the early morning of 25th September, when a bird of the year was killed at the lantern.

REGULUS REGULUS, *Goldcrest.*—The first Goldcrests appeared on board singly during the forenoon of 23rd September, others again on the 29th and on 1st October, when they were moving southwards during the daytime. Numbers were flying in the rays, and fluttering against the lantern during the early morning movements of 29th

September and 2nd and 3rd October, and on the night of 17th to 18th October.

PHYLLOSCOPUS COLLYBITA, *Chiffchaff.*—At the lantern with other species on 3rd October. One was captured at 3 A.M.

PHYLLOSCOPUS TROCHILUS, *Willow-Warbler.* — Was moving southwards during the daytime on 19th, 20th, and 22nd September; and on the early morning of the 25th was flying around the light, when several examples were killed. Came on board freely, but singly, in the daytime, and on leaving went south-south-west.

HYPOLAIS ICTERINA, *Icterine Warbler.*—One came on board on the afternoon of 22nd September, and allowed an inspection at close quarters before it quitted the ship for the coast of Kent.

MOTACILLA LUGUBRIS, *Pied Wagtail.*—A few were moving to the south-south-west, during the daytime, late in September; but the bird was chiefly observed migrating to the south-east and south-south-east, towards the coast of the Continent, between daylight and 10.30 A.M. on 1st, 7th, 14th, and 16th October.

MOTACILLA FLAVA, *Blue-headed Yellow Wagtail.*—A fine adult male, in newly assumed winter plumage, was captured on the ship at 3.30 P.M. on 22nd September, a great date for diurnal movements.

MOTACILLA RAII, *Yellow Wagtail.*—One alighted on the deck at 1 P.M. on 22nd September, remained a few moments, and then flew south-south-west. Many other species were moving southwards at the time, and probably other representatives of this species.

ANTHUS PRATENSIS, *Meadow-Pipit.* — This, one of the most frequent migrants, was observed passing on seventeen days. The main line of flight for emigrants was from north-west to south-east (from the Essex coast towards that of Belgium), and considerable numbers passed in that direction from 18th September to 14th October between 6 A.M. and noon. Smaller numbers were observed moving to the south-south-west. The immigrants came from both the east and south-east, chiefly from the former quarter, and passed in numbers towards the Essex coast, from 7th to 16th October, the chief flights being on the 8th, when some of the parties were forty strong. Was present during the night movements of 3rd and 18th October, but in small numbers, a few being killed.

ANTHUS TRIVIALIS, *Tree-Pipit.*—Was moving southwards during the daytime between 19th and 23rd September, when several individuals came on board singly, some of which were captured. In the early morning of the 25th one was killed at the lantern along with other emigrant summer birds.

ANTHUS RICHARDI, *Richard's Pipit.*—A male, in first plumage, was captured at the lantern at 3 A.M. on 25th September, during a considerable movement of summer birds, and, no doubt, came from the east.

ANTHUS OBSCURUS, *Rock-Pipit.*—Rock-Pipits, probably moving southwards, came on board on 23rd September and 12th October during the daytime, and one was captured at the lantern at 6.45 P.M. on 8th October.

MUSCICAPA GRISOLA, *Spotted Flycatcher.*—Was flying around the ship, with a number of other species, from 1 to 4 A.M. on 25th September, and two were killed at the lantern.

MUSCICAPA ATRICAPILLA, *Pied Flycatcher.*—Was first observed, and captured, on the night of 18th September;

some alighted on the ship on the following day, when they proved their expertness in both finding and capturing insects, a "gamma" moth not being too much for them. On the 29th one was taken off the lantern at 3 A.M.

HIRUNDO RUSTICA, *Swallow.*—Swallows, old and young, were flying together to the south-south-west on 26th September, and 2nd, 13th, 14th, and 16th October; and to the west on 26th September and 14th October. The chief movements were on 13th and 14th October, when parties, some of them one hundred strong, passed southwards, all flying low over the surface of the water. On the latter date, they were observed passing both westwards and southwards—*i.e.*, making for and departing from the English coast! The earliest hour for these migrants was 7.45 A.M., and the latest 2 P.M.

CHELIDON URBICA, *House-Martin.*—The first and greatest movement of this bird was observed on 9th October, when, after a decided fall in temperature, numbers were passing from north-west to south-east, from 9 A.M. to 1 P.M., some of the parties containing as many as fifty individuals. Smaller numbers passed in the same direction on the 13th. On the 14th and 16th small flocks were moving from east to west in the forenoon. On the 16th several went north-west during the morning.

COTILE RIPARIA, *Sand-Martin.*—On 15th October the watch reported that he had seen twenty "swallows" passing south-south-west at 5.45 A.M., and of these, seven alighted on the rail and rested for ten minutes. They were quite tame and allowed a close approach, and were described as being "brown above and white below."

PASSER MONTANUS, *Tree-Sparrow.*—From 23rd

September until 8th October single birds or pairs came on board from the east at intervals. On the last-named date the bird passed in considerable numbers going due west, and again on the 11th and 18th. On the 11th a small party was observed flying to the north-west. Was not seen at night. This species came on board more frequently than any other, and displayed many of the traits characteristic of its commoner cousin, being very noisy, and having the knack of making itself at home, even at sea. When aboard during high winds and heavy rain, the birds used to hustle each other in the scramble for the most sheltered places in the rigging and on the lee side of the lantern, and showed much pugnacity.

FRINGILLA CŒLEBS, *Chaffinch.* — First seen on 29th September, when an adult male came on board at 5 P.M. Not observed again until 8th October, when the great east-to-west flights set in, in all of which this bird participated largely, passing in flocks in the fore- and afternoon. On the morning of 15th October, it was passing in flocks to the south-south-west as a British emigrant, and to the west as an immigrant. Many were present during the great night movement of 17th to 18th October.

STURNUS VULGARIS, *Starling.* — First observed on 24th September, but down to 8th October single birds only were seen, passing occasionally to both the south-south-west and west during the daytime. On 8th October small parties passed from east to west, and this, the first decided movement in this direction, was followed by others of considerable magnitude, which have already been described. Small numbers were occasionally observed emigrating from north-west to south-east during

October. A solitary individual only came under notice at the lantern during the important nocturnal movements which took place in the latter half of September. The second occasion on which this species was observed at the light was during the great night movement of 17th to 18th October, when thirty-two examples were killed or captured, all of them of the ordinary green-headed race. I secured only one specimen of the purple-headed form, and this came from the east on the afternoon of 28th September. I much regretted not being able to ascertain to what race the vast numbers passing from east to west belonged, for not a single bird came aboard during the larger movements. It was impossible to say from what quarter the birds taken during the night movement alluded to came. Some of my friends regard the purple-headed birds procured by me at the Eddystone as merely fresh-moulted specimens of the ordinary bird. If this be so, how is it that all the fresh-moulted examples obtained at the Kentish Knock at an almost identical date had green heads?

CORVUS MONEDULA, *Jackdaw.*—A few appeared from the east at 4.25 P.M. on 17th October, and others followed and were flying round the ship until 5 A.M. on the 18th. Two were captured.

CORVUS FRUGILEGUS, *Rook.*—The first Rooks appeared in small numbers from the east at 5 P.M. on 17th October, being preceded by a few Jackdaws. Later several appeared at the lantern, and flew in the rays of light from 7 P.M. to 3 A.M. on the 18th, an adult being captured.

ALAUDA ARVENSIS, *Skylark.*—This species was more in evidence than any other, and its movements were of a singularly varied nature. As an emigrant from Britain it was observed moving to the south-south-west, south, and south-east; and as an immigrant, to the west and north-west. It participated largely in the southerly migrations, both by day and night, during the latter half of September. It was also the most frequent visitor to the light, and was never absent from any of the night movements. As with other species, its pronounced flights from east to west were not observed until 8th October, but after that date it took a prominent part in all the great westerly flights. It was present in numbers during the great night migration of 17th to 18th October. Many were killed or captured at the lantern.

FALCO TINNUNCULUS, *Kestrel.*—An adult male flew in the rays and approached the lantern continually between 8 P.M. of 18th September and 1.30 A.M. on the 19th.

PHALACROCORAX CARBO, *Cormorant.* — Single birds were seen on 18th and 21st September.

SULA BASSANA, *Gannet.* — Not unfrequently seen, moving chiefly southwards, and always in adult plumage. Did not fish in the vicinity of the lightship.

ŒDEMIA NIGRA, *Common Scoter.* — First seen on 28th September, and frequently afterwards, usually flying towards the feeding grounds on the fringe of the Kentish Knock sands.

ÆGIALITIS HIATICOLA, *Ringed Plover.*—This species was heard on five occasions, between the hours of 6.40 P.M. and 1.15 A.M., passing over the ship towards the west or north-west—namely, on 17th, 19th, 24th, 25th, and 29th September. The unknown notes of other Limicolæ were, on three of these occasions, heard at the same time.

VANELLUS VANELLUS, *Lapwing.*—Between 9.45 P.M. and 11 P.M. on 30th September, during moonlight,

Lapwings were heard passing overhead from east to west.

PHALAROPUS HYPERBOREUS, *Red-necked Phalarope.*—At midday on 30th September, one was seeking food on the water at some little distance from the ship.

TOTANUS HYPOLEUCUS, *Common Sandpiper.*—Heard passing southwards at 10.45 P.M. during the night movement of 18th September, when Redstarts and Pied Flycatchers were flying around the lantern.

STERNA FLUVIATILIS, *Common Tern.*—An immature example appeared and alighted on the rail during the forenoon of 29th September.

RISSA TRIDACTYLA, *Kittiwake.*—Adults and young were common from 22nd September onwards.

LARUS ARGENTATUS, *Herring-Gull.*—An adult seen on 15th October was the only example of this species observed.

LARUS FUSCUS, *Lesser Black-backed Gull.*—Seen almost daily, but in greater numbers after 2nd October.

LARUS MARINUS, *Greater Black-backed Gull.*—A few seen daily after 3rd October. I never noticed any decided movements on the part of gulls.

STERCORARIUS POMATORHINUS, *Pomatorhine Skua.*—From 7th October was seen daily in attendance upon the Lesser Black-backed Gulls and Kittiwakes. Few mature birds were seen.

STERCORARIUS CREPIDATUS, *Arctic Skua.*—Present daily from 24th September onwards, chiefly engaged in bullying the Kittiwakes and sometimes the Lesser Black-backed Gulls. The dark form largely preponderated.

FULMARUS GLACIALIS, *Fulmar.*—On 2nd October one was observed flying southwards at 5.45 P.M.

COLYMBUS SEPTENTRIONALIS, *Red-throated Diver.*—

Both adults and young were seen not unfrequently from 22nd September onwards. They were always fishing singly near the ship.

URIA TROILE, *Common Guillemot.*—Not uncommon off the edge of the sand, where the shallow water probably affords good fishing ground. These birds, and others seeking similar situations, were most numerous during strong westerly winds, when the east side of the sand afforded shelter and the possibility of obtaining food.

ALCA TORDA, *Razorbill.*—The same remarks apply to this species as to the last.

FRATERCULA ARCTICA, *Puffin.*—A single bird seen on the wing on 10th October was the only record.

The following additional species have also been observed at this lightship. Some of them were sent to me after I left the vessel, while the records of the rest have been culled from the schedules of observations furnished to the British Association's Committee from 1880 to 1890.

FRINGILLA MONTIFRINGILLA, *Brambling.*—The wings of six Bramblings killed at the lantern on 9th October 1887 were sent to Mr Cordeaux. Hundreds of birds were present at the light on this occasion.

PARUS MAJOR, *Great Titmouse.*[1]—Five on deck at midday on 26th November 1903. One of these, an adult female, allowed itself to be captured, and was sent to me. "All the birds were of the same species, and uttered a note like a Chaffinch" (G. E. Highton).

SYLVIA BORIN, *Garden Warbler.*—One struck the lantern on the morning of 4th May 1904, and was sent to me.

[1] Probably the typical Continental race.

ERITHACUS RUBECULA, *Redbreast.*—There are five records in the schedules for the occurrence of this familiar bird. Only one of these is for spring—namely, for 17th April 1888, when a single bird was on deck at 6 A.M. The autumn records are for 21st October, and 8th, 10th, and 12th November. On 10th November, fifty or sixty were at the lantern from 2 A.M. to daylight.

TROGLODYTES TROGLODYTES, *Wren.*—Two appeared on board at 4 P.M. on 13th October 1885, and roosted in the reefed sail.

ALCEDO ISPIDA, *Kingfisher.*—One found on deck at 10 P.M. on 20th March 1904 was forwarded to me.

CUCULUS CANORUS, *Cuckoo.*—One which struck the lantern at 5 A.M. on 10th May 1904 was sent to me.

SCOLOPAX RUSTICULA, *Woodcock.*—There are only two records of the visits of this well-known migrant. On 30th October 1885, one was killed at the lantern at 11.30 P.M.; and another on 24th October 1890, at 2 A.M.

TRINGA ALPINA, *Dunlin.*—On 20th March 1904, one was found on deck, along with the Kingfisher, and sent to me.

NUMENIUS ARQUATA, *Curlew.*—Two occurrences only are recorded. On 6th August 1880, one was seen at 11 A.M.; and on 30th April 1888, many were at the lantern at 9.30 P.M., and one struck and was killed.

PROCELLARIA PELAGICA, *Storm Petrel.*—A few were at the lantern from 6.30 P.M. to 12 A.M. on 11th November 1885.

NOTE.—The sequence of the species of Passerine birds in this chapter is not quite the same as that followed in the rest of the studies. It is that of a paper which appeared in *The Ibis* for 1904.

[*Photo: C. Dick Peddie.*

FAIR ISLE FROM THE SOUTH-WEST.

CHAPTER XIX

FAIR ISLE, THE BRITISH HELIGOLAND

On consulting a map of Scotland, with a view to selecting a bird-watching station in which to spend my autumn vacation in 1905, I was much impressed with the favourable situation of Fair Isle for that purpose. It seemed, theoretically, to afford quite a number of exceptional advantages. It appeared to me to lie right in the line of flight of the hosts of migrants which in spring and autumn traverse our shores and islands when proceeding to and from their summer homes in Northern Europe. Another important feature was its isolation, which would render it a most welcome resting-place for the migrants performing these seasonal passages. Here, too, the bird visitors would be concentrated within narrow limits, and its moderate size would make it possible to ascertain, with some degree of accuracy, what species were present daily. All these factors are of extreme importance to the would-be observer of bird-migration.

Besides these considerations there was yet another—namely, the fact that practically nothing was known concerning the bird-life of this little-visited island, though it may be noted that its falcons were famous some three hundred years ago.

My forecast of the importance of Fair Isle as a

bird-observatory has been more than realised. Seven years' investigations have made it the most famous bird-observatory in our islands; indeed it has become the British Heligoland. Quite a number of species which were previously regarded as rare casual visitors to our isles, have, as the result of these observations, been found to be regular migrants. They doubtless occur on the mainland, too, but owing to its broad acres they almost entirely escape notice. In addition, several species have been added to the British and many to the Scottish avifauna through the Fair Isle investigations.

Though lying midway between the Orkney and Shetland groups, Fair Isle has remained among the least visited of all the numerous inhabited islands in the British seas. This is, no doubt, to be accounted for by there being no regular communication by steamer with the island, a fact which is probably due to the entire absence of a reasonably good natural harbour, and to the fierce tidal streams which rush along its rugged and precipitous coasts.

The island is situated some 24 miles south-south-west from Sumburgh Head, the southernmost point of the mainland of Shetland, and 26 miles east-north-east of North Ronaldshay, the most northerly of the Orkneys. It is somewhat oblong in form, having an extreme length of about 2¾ miles from north to south, and averaging about 1 mile in breadth. Though a circuit of the island may be made by a walk of 9 miles, yet so indented and irregular is the coastline that its in-and-out circumference is not much short of 20 miles.

Except a small portion of its southern coast, and a little inlet on the east, the island is everywhere surrounded by a belt of precipitous cliffs, ranging from

[*Photo: W. Norrie.*

FAIR ISLE: THE NORTH-WEST CLIFFS.

100 to close upon 600 feet in height. At intervals this lofty coastline juts out into remarkable peninsulas and bluff headlands; and its face has been scooped out to form great picturesque geos and innumerable caverns; while many natural arches, detached pillars of rock, lofty stacks, and skerries add much to the grandeur and interest of the coast scenery. This great belt of natural precipice—the resort of hosts of sea-fowl in the summer—is highest on the west side; and though the cliffs on the east are not so lofty, yet here they are graced by the picturesque "Sheep Craig"—a noble mass of rock rising almost perpendicularly from the sea to a height of some 500 feet, which is one of the greatest resorts of sea-fowl in the island, and formerly afforded a fitting home for the Sea Eagle.

Seen from the south harbour, the appearance of the isle is decidedly picturesque and "fair." In the centre of the foreground are the crofts, about 200 to 300 acres in extent, and all golden and green at harvest time. These are flanked on either side by high grass-clad ground, and a fine skyline is formed by the singularly irregular outline of the cliffs, and the undulating contour of the high ground on the north.

The northern two-thirds of the island are mostly barren, being either carpeted with stunted heather, grass, and a creeping species of juniper, or bare and stony, the turf having been torn off for fuel. The ground here is high, especially on the west, and culminates in the Ward Hill (712 feet), the highest point in the island. On the lower ground there is an extensive wet area known as "Sukka Moor," with a number of small lochans, and traversed by a little burn. This portion of the island would seem to be well suited as a nesting-haunt for the Golden Plover, Curlew, Whimbrel, Snipe, Lapwing, and Dunlin, but the presence of the descendants of the once famous Peregrines probably accounts for the absence of these waders as breeding species, and now this fastness is sought only by Wild Geese as a safe retreat during their short sojourns, and as a resting-place for moulting Gulls in the late summer.

There are two natural harbours. Of these, that on the south is the one used, weather permitting. It is, however, beset by numerous submerged rocks, and across its mouth rushes a furious tidal roost, making it necessary to have native experience ere one ventures to enter it in a boat. There is a smaller and better harbour on the northern section of the east coast, but it is somewhat removed from the inhabited portion of the island.

There are several small burns which, after a short course, discharge themselves into the sea as waterfalls down the face of the cliffs; and there are two small sheets of water at the north end of the island.

There are no trees or shrubs of any description, either native or cultivated, and thistles and bracken, the latter chiefly confined to a small belt on the north-east side, are the giants of the indigenous flora.

The natives number about 130. They are crofter fishermen, and, though well-housed, live in the same primitive manner as their forefathers. They have always been most kindly disposed towards me, and have afforded me every facility in their power for carrying out my investigations. That Fair Isle has been inhabited for many centuries is manifest from the presence of tumuli, in which cinerary urns have been discovered.

Since 1892, there have been two lighthouses on the island—namely, at Scaddon, at the extreme south-west,

[*Photo: W. Norrie.*

FAIR ISLE: A RIFT IN THE WESTERN CLIFFS.

and at Skroo, at the north-east limit. Both are furnished with powerful white revolving lights, the beams of which are arranged into groups.

The island is hardly known to the general public save perhaps as the scene of the wreck, in the autumn of 1588, of "El Gran Grifon," one of the ships of the Spanish Armada, whose crew spent several months there in a more or less starving condition, and in great wretchedness, for the dwellings of the inhabitants were then the filthiest of hovels, and the natives poverty-stricken in the extreme. It can boast, however, of having received some distinguished visitors in the past, for Sir Walter Scott landed there on 14th August 1814, and spent several hours on the island; and Mr R. L. Stevenson paid a short visit on 21st June 1869.

I have visited Fair Isle for five consecutive autumns, 1905-1909, remaining for five weeks on each occasion; and I made three visits in spring—namely, in 1909, 1910, and 1911. On my first two visits I had the valuable co-operation of Mr Norman B. Kinnear, and since then George Stout, a Fair Islander, who had been trained by Mr Kinnear and myself, has rendered me great service. During the springs and autumns of 1909, 1910, and 1911, Her Grace the Duchess of Bedford added materially to the results obtained.

In the year 1908 the investigations entered upon a new era. Being then convinced that the island was a most important station for observing the movements of migratory birds, I determined, if the necessary help were forthcoming, to obtain a day-to-day record of its feathered visitors; to appoint, in fact, an observer whose whole time would be devoted to the investigations. Thanks to the generosity of a few friends, I was enabled to carry out my project, and George Stout was appointed recorder. Through his assiduity and excellent work, most satisfactory results were obtained. In 1909 George Stout left the island, and his brother Stewart, also an enthusiastic bird-watcher, took up the work with success. Since January 1910 Jerome Wilson has proved himself to be an assiduous and careful observer.

As the result of six and a half years' investigations, this insignificant island has been visited by no less than 207 species, or *about one-half of the birds that have ever been known to have occurred in the British Isles!* Nor are its resources in this respect by any means exhausted; on the contrary, each year adds its quota of important records and novelties; and it is intended to continue the investigations for some time to come.

The outstanding feature of its bird-life is the importance of the passage movements, for the observation of which it is not only unrivalled as a British station, but has few equals anywhere. Extraordinary numbers of these migrants appear regularly during the spring, when on their way to, and in the autumn when returning from, their wide and far-extending nesting-grounds in Northern Europe, Iceland, and Western Siberia. The knowledge gained from the Fair Isle statistics has thrown a flood of light upon these important and in some respects obscure migrations, such as was never before possessed for the British Islands. It has been ascertained with a surprising degree of accuracy what species participate regularly in these great movements, and the dates between which they are performed at both seasons. It has been possible, also, to note the increase in the stream of migrants under incentives highly favourable for their performance, its

[Photo: N. B. Kinnear.

FAIR ISLE: THE SHEEP CRAIG FROM THE SOUTH-WEST.
(A former breeding-place of the Sea Eagle.)

[Photo: C. Dick Peddie.

FAIR ISLE: THE SOUTH-WEST CORNER AND SKADAN LIGHTHOUSE.

arrest during stressful periods, or, again, its even flow under ordinary conditions: in other words, the island has afforded the opportunity of correlating the divers movements with the weather conditions, and ascertaining what the meteorological incentives, checks, and barriers, as the case may be, to migration are—a knowledge of the relations existing between the two sets of phenomena which was highly desirable.

This superiority of Fair Isle for these important researches arises from its isolation and its small size; otherwise it has no advantages over the other Isles of the Shetland and Orkney groups. In both these archipelagos, however, the islands are many, not a few of them are large, none are far apart, and hence the migrants visiting them are widely and thinly scattered, and in this way the great majority of their bird-visitors entirely escape notice. Fair Isle itself, with its 2½ square miles of varied surface, and its extensive belt of lofty cliffs, is too spacious even for several observers; and there were many days, when migrants were abundant, that we were conscious that in spite of strenuous and unremitting endeavours much had been missed—indeed, that the great majority of the visitors had not come under notice.

The experience gained during many vacations spent in some of the most favoured observing-stations in the British Isles and elsewhere, has convinced me that we see an infinitesimal portion of the migrants which visit our shores. This is especially the case on the mainland, with its vast extent of coastline, its enormous acreage of enclosed ground, and its extensive woodlands and other forms of cover. On reaching the mainland, the migrants, particularly the Passerine birds, seek suitable haunts where cover abounds, and thus few—very, very few—come under notice. We should remember, too, that we have not in the British Isles, where bird-watchers are more numerous than elsewhere, anything like one daily observer for every 100 square miles of country! Small wonder, then, that so very much escapes notice. It is well to bear this in mind when drawing deductions from migration data covering large areas.

To return to Fair Isle: it must not be supposed after a great immigration overnight, that birds will be in evidence everywhere on the following day. This only applies to species that show a predilection for the open country, such as Fieldfares, Wheatears, Pipits, and the like. The various species of Warblers, the Bluethroat, Thrush, Blackbird—indeed, the majority of the arrivals (I allude to the autumn)—are either in hiding among the turnips, potatoes, and standing corn of the crofts, or are quite beyond the range of observation on the face of the great cliffs.

The crofts are the great hunting-grounds, and fortunately my numerous friends among the islanders have most kindly permitted me to search their cultivated ground. If this great privilege had been denied, the labours of myself and my coadjutors would have been well-nigh in vain. The crofts require to be searched in a most thorough and systematic manner, for the birds, being more or less exhausted by their long flight overnight, lie very close among the variety of cover they afford and are not easily discovered. When disturbed, the migrants only fly a few yards ere they drop into cover again—a trait which makes their identification a matter of great difficulty, for one only gets a hurried glimpse of them, and this, too, amid very unusual

[*Photo: Duchess of Bedford.*

FAIR ISLE: PART OF THE CROFTED AREA.

surroundings. However, in time one becomes more or less familiar with most species, and readily detects a stranger—but only as a stranger, until it has been brought to hand: one must shoot in such cases; if not, the identity of some of the visitors would remain a mystery. Those who have not engaged in this kind of bird-work have little idea how puzzling it is to identify common species under such very unusual conditions. Indeed, quite familiar species are not immediately recognised, unless they possess some very marked diagnostic character. The immigrants, too, are, with few exceptions, such as the Thrushes, remarkably silent. During the spring movements, however, I have on several occasions heard some of the travelling birds (the Willow-Warbler, Whinchat, and Ring-Ouzel) indulge in a few notes of a very subdued song—one or two birds only out of thousands.

The great cliffs, more particularly the western range, are, alas, also a great resort of the smaller bird-travellers, and of such rock-loving species as the Ring-Ouzel. The reflected heat of the sun renders these haunts particularly genial, and their faces, abundantly clad with lichens, are alive with insects on fine days. Here the insectivorous species swarm after a great arrival of migrants, Redstarts, Flycatchers, Warblers, Goldcrests, Tree-Pipits, Hedge-Accentors, and Redbreasts being much in evidence. The most remarkable visitor to the face of these cliffs is the Woodcock. In autumn not a few of these birds may be seen, where there are grassy ledges, resting in their usual posture, with their tails up and their bills down. I said "alas!" the reason for my lament being that these fastnesses are not only vast, but are almost entirely unapproachable, and hence all but impossible for observation. On this account thousands of the migrants which resort to these miles of precipitous cliffs entirely baffle the efforts of the most assiduous and venturesome of watchers. The only places where one can get a peep at them are on the very limited portion of the cliffs which flank the geos, but even there one sees the merest fraction of those present. What one does witness, however, is of great interest, for it demonstrates the hardships incurred, and the shifts for a living that have to be resorted to by birds during their migratory flights. On these rocky fastnesses, Goldcrests may be watched creeping in numbers on the faces of the gaunt lichen-spangled precipices in the pursuit of food, and, if near enough, may be heard uttering the while their feeble notes, which strike one as savouring of irritability —perhaps a natural reflection of their feelings on finding themselves committed to such very unusual hunting-grounds. Redstarts and Flycatchers (Common, Pied, and rarely Red-breasted), Wheatears, Pipits (Meadow and Tree), hitherto unseen, may be observed darting out from these retreats to capture some insect on the wing; and Thrushes, Redwings, Blackbirds, and Ring-Ouzels seen hopping about in search for what they can find (there are no mollusca), to allay the cravings of hunger.

While the corn (bere and oats) is standing, it harbours various species of Warblers, and is the favourite retreat of the Ortolan Bunting. When it is cut, the stubbles afford suitable and much-frequented haunts for the different species of Finches and Buntings, and for the ubiquitous Twite, which is to be found there in thousands. Associating with the latter, most unfortunately, are several other species of the finch family, such as the Little Bunting; these are most difficult, often quite impossible, to detect amid the restless, noisy crowd of undesirables, from an investigator's point of view.

The high ground is the chief resort, during the migratory period, of the Woodcock and Fieldfare. The former is sometimes extremely abundant there on the days following its arrival overnight, every stone and tussock being resorted to for concealment. Very few Woodcock are seen on the lower ground and in the crofts.

In spring there is practically no cover to be found in the cultivated portion of the island, and the numerous passage migrants resort to the sides of the burns and drains, and other places where there is rough grass. The Finches and Buntings resort to the cultivated ground where corn has been newly sown. On this bare land during the spring rush of migrants on their way north, it is a strange spectacle to see such species as Pied and Spotted Flycatchers, Redstarts, Whinchats, Reed-Buntings, Tree-Pipits, and even Wrynecks, in search of food. The cliffs at this season are the main resting-places, and are much resorted to by all kinds of migrants. The island does not look its best in spring; indeed, it offers a great contrast to the attractive appearance it wears in the autumn.

Migrants on arriving go into hiding, to rest after their overnight journey, and remain concealed until about midday, when they come out to search for food. In the autumn some of them may be sought for in the potatoes and turnips; in like manner in spring they are not to be found early in the day, for they lie hidden in places where it is impossible to find them, as they also do during the time of high winds and heavy rain at both seasons.

In autumn the Warblers, Flycatchers, and other non-gregarious birds are very silent, and do not utter any note when disturbed among cover. Sometimes, however, one hears the pretty call-note of the Willow-Warbler, and the low plaintive note of the Yellow-browed Warbler, but I do not remember to have heard any other. Though these birds may be numerous they are only found singly, and not in company or in parties.

It is otherwise with the Finches and Buntings. These birds continually "call," even when in parties, but particularly so when met with singly, as if to get a response from another of their kind. The Blackbirds, Ring-Ouzels, and Fieldfares are noisy when alarmed; the Thrushes and Redwings are less so.

Fair Isle is richest in the number of its Passerine visitors, and poorest in those representing the Waders and Ducks. Its rock-bound coastline, with scarcely a break in it for the formation of a beach, offers little or no attraction for shore-birds. The reefs at the south-western corner of the island are the chief resort of the waders that do alight; but this haunt is a most difficult one to explore, owing to its remarkably rugged nature, due to the outcrop of the strata being almost perpendicular, and their edges extremely irregular, sharp, and saw-like—a combination which renders investigation both a painful and a rough-and-tumble process, during which most of the birds slip away unnoticed. The Isle is also deficient in congenial haunts for the various kinds of ducks which obtain their food in fresh water, especially for the diving forms, and the few of the latter which have occurred have generally appeared singly.

For the marine forms, the Isle, being situated amid a stormy main, affords little shelter, and hence it is chiefly such hardy seagoing species as the Eider and the Long-tailed Duck that can brave the turbulent Fair Isle waters and find in them haunts congenial to their tastes. The surface-feeders, such as the Mallard, Teal, and Wigeon, are, however, to be found in small numbers, during the autumn, winter, and spring, on the few freshwater pools, and on the sea at the mouths of the burns.

The numerous migrants arrive almost entirely during the hours of darkness, and after a longer or shorter sojourn, depending on the state of the weather, take their departure during the night, both their incoming and outgoing being usually unobserved. An interesting exception to this rule came under notice on the evening of 9th October 1908. During this day there were thousands of Redwings on the island which had arrived during the previous night, and at 6 P.M. a large party of these birds were observed to rise high on the wing and leave the island, proceeding in a south-westerly direction towards North Ronaldshay — the northernmost island of the Orkneys.

On fine sunny days during September, I have frequently seen Skylarks and Pipits leaving the island between 6 A.M. and noon. They usually left in parties (some of the Skylarks flying high), and these also shaped their course for the Orkneys.

Fair Isle has produced a surprising crop of rare species—more so than any other portion of the British area during the period covered by the investigations, 1905-1911—and several of them have proved to be new to the British fauna and quite a number to that of Scotland. The occurrences of rare birds have always had a peculiar charm for ornithologists; but to the student of bird-migration, their irregular, and in many instances presumably erratic visits, are the very reverse of helpful in connection with his researches. The appearance of Central, Eastern, and Southern European species far beyond their wonted homes, and quite off their regular accustomed lines of flight during migration, is distracting, and presents problems which do not admit of satisfactory explanation. All that we can opine is that such occurrences are possibly due to some unaccountable failure of that special faculty already alluded to, possessed by migratory birds, which leads them unconsciously to the particular seasonal haunts they should seek. The stimulus to migrate has certainly been strong within the wanderers, or many of them would never have reached such a far-off goal as Fair Isle.

The following species obtained at Fair Isle are new to British fauna:—Pine Bunting, Thrush-Nightingale, Northern Willow-Warbler, Blyth's Reed-Warbler, and Red-rumped Swallow. Among the other birds of extreme rarity as visitors to the British Isles are the Black-throated Wheatear, Subalpine Warbler, Siberian Chiff-chaff, Lanceolated Grasshopper-Warbler, Savi's Warbler, Red-throated Pipit, Greenland Redpoll, Black-headed Bunting, Rustic Bunting, Yellowshank, etc.

The discovery of the regular occurrence of such former rarities as the Yellow-browed Warbler, Red-spotted Bluethroat, Little Bunting, Ortolan, and Grey-headed Wagtail, and the not infrequent visits to Fair Isle of other species which were once supposed to be quite irregular in their appearance in the British Islands, has resulted in the necessity for their status as British species being modified, since they can no longer be classed as "rare casual visitors," which was formerly their designation as members of our avifauna.

The Sea Eagle is one of the birds of the past as a native species. It formerly had an eyrie on the Sheep Craig, but was banished sometime between the years 1825 and 1840. There is another site of a former eyrie, but this was probably an alternative nesting-place of the birds alluded to, and not the domain of a second pair.

The question as to whether young and old birds of the same species migrate in company is an interesting one, and perhaps, as bearing upon it, the following short list of birds which I have observed at Fair Isle so doing, may be worth giving:—

Rook, Starling, Siskin, Snow-Bunting, Wheatear, Ring-Ouzel, Bluethroat, Brambling, White Wagtail, Lapland Bunting, Swallow, Martin, Blackcap, Redstart, Chaffinch, Wigeon, Mallard, Turnstone, Golden Plover, Lapwing, Dunlin, Black-headed Gull, Common Gull, and Iceland and Glaucous Gulls.

A list of all the native and migratory birds known to Fair Isle, with an indication of their times of arrival, departure, and passage, will be found in a subsequent chapter.

CHAPTER XX

A YEAR WITH THE MIGRATORY BIRDS AT FAIR ISLE

THIS chapter is based upon the records contained in the diaries of George Stout and myself for the year 1908. Its aim is to furnish a chronological account of the movements witnessed during an entire year at a famous station. Such a record is useful, since it affords interesting and important information as to the dates covered by the migrations of the various species; the birds travelling in company; the habits of the voyagers when en route; the weather associated with the movements and its influence as an incentive to their performance, or as a deterrent to their progress; and other points of interest.

In connection with the weather, it has been thought best to give that recorded at Fair Isle itself, since the types of weather in the British Isles, and their influences on bird-migration, have been specially treated of in Vol. I., pp. 171-187.

The year 1908 has been chosen because it was George Stout's last on the island, and hence we have his best work. My contributions were made during a five-weeks' sojourn in September and early October.

1908

1st January.—It will be well to commence the year with a short account of the birds which pass the winter in the island.

I. The following are RESIDENT species, and most of them are probably represented by the same individuals all the year round :—

RAVEN.	TWITE.	EIDER DUCK.
HOODED CROW.	ROCK-PIPIT.	GREAT BLACK-BACKED GULL.
STARLING.	WREN.	HERRING GULL.
HOUSE-SPARROW.	PEREGRINE FALCON.	BLACK GUILLEMOT.
TREE-SPARROW.	SHAG.	

The Kittiwake, Common Guillemot, and Razorbill breed and during the winter are seen at sea off the island.

II. The REGULAR WINTER VISITORS are :—

SKYLARK.	FIELDFARE.	SNIPE.
CHAFFINCH.	BLACKBIRD.	PURPLE SANDPIPER.
BRAMBLING.	CORMORANT.	REDSHANK.
SNOW-BUNTING.	LONG-TAILED DUCK.	GLAUCOUS GULL.
REDBREAST.	MALLARD.	COMMON GULL.
SONG-THRUSH.	WIGEON.	LITTLE AUK.

III. The following are OCCASIONAL WINTER VISITORS :—

GREENFINCH.	GANNET.	BLACK-HEADED GULL.
MEALY REDPOLL.	TEAL.	ICELAND GULL.
CORN-BUNTING.	WATERHEN.	PUFFIN.
REDWING.	LAPWING.	LITTLE GREBE.
KESTREL.	OYSTER-CATCHER.	FULMAR.
HERON.	CURLEW.	

8th January.—Northerly breeze ; clear.

Black-tailed Godwit, one shot on a flooded croft. Several Rooks on the island.

9th January.—Northerly breeze; clear.

Lapwings, seven, arrivals. About a dozen Song-Thrushes also present.

13th January.—South-west breeze; clear.

More Lapwings, over a dozen. Three Rooks, arrivals. A small party of Mealy Redpolls seen.

20th January.—Guillemots and Razorbills visiting the breeding-ledges on the Sheep Craig for the first time this season.

28th January.—Two Mealy Redpolls observed among a flock of Twites. The Fulmars now very numerous. They began to come in about the first of the month, and now there are scores on the face of the Sheep Craig alone. A Curlew noted.

4th February.—Two Glaucous Gulls on land. This species is fairly common in winter, and in the stormy weather resorts to the crofts, along with the commoner Gulls. The Guillemots and Razorbills are beginning to visit the cliffs in numbers. Large numbers of Eider Ducks, males and females, are now appearing. Through the winter there are always a few females to be seen in the geos, but the old males are always in the minority, in the months of December and January especially. The numbers of Eiders to be seen round the Isle now are at least double those which nest. In late autumn large flocks constantly arrive and pass on, and it has been noticed that they are most numerous after a strong northerly wind.

8th February.—After a westerly breeze, numbers of Glaucous Gulls present. At least thirty in one of the sheltered bights on the east; and in their company, four Iceland Gulls, two adult and two young.

10th February.—First Oyster-catcher for the season

observed. Shag with nest nearly complete, and several more building.

14th February.—Two Meadow-Pipits, the first for the season, noted to-day. Also a Waterhen, which is a somewhat rare visitor.

17th February.—The Rooks which have been here for some time departed.

29th February.—During last week, numbers of Skylarks were observed arriving, and now they amount to considerable flocks. To-day three Curlews and one Merlin seen.

3rd March.—A Brambling seen among Chaffinches. A flock of Oyster-catchers, numbering about thirty, have arrived; they may possibly be our summer residents.

4th March.—Two Golden Plovers.

5th March.—Two Lapwings, fresh arrivals.

7th March.—Wind south-east; clear.

Over six hundred Lapwings have put in an appearance to-day. Parties of Starlings in company with them are also undoubtedly immigrants. A Glaucous Gull, a Ringed Plover, and several Fieldfares seen; latter may be the wintering birds.

8th March.—One Heron noted as an arrival.

9th March.—Two Long-tailed Ducks, male and female. Meadow-Pipits, two more arrivals. A Pied Wagtail appeared;, also three Golden Plovers. Lapwings decreasing in numbers.

10th March.—Easterly breeze; clear.

The new arrivals to-day are a Wood-Pigeon and another Pied Wagtail. Several Redbreasts observed, some of which may be immigrants. Skylarks and Lapwings numerous, but the latter decreasing. The Kittiwakes in numbers at the Sheep Craig for the first time this year. Three Mallards and several Glaucous Gulls also seen.

11th March.—Wind changed to north-east last night.

Two Ringed Plovers are new arrivals. Several Fieldfares have appeared since yesterday. The Lapwings are now reduced to two hundred birds.

12th March.—Same birds as yesterday, with the first Dunlin of the season.

14th March.—South breeze, A.M.; south-west, P.M.

Over a dozen Rooks appeared this morning. One Redpoll and a Gannet were also seen.

15th March.—Three Pied Wagtails to-day.

16th March.—First Corn-Bunting, also a Pied Wagtail.

17th March.—South wind continues.

Two Lesser Black-backed Gulls, first of season. A male and female Merganser, a male Stonechat, a Mallard, and Wigeon also noted. Very few Lapwings now.

19th March.—Two male Stonechats and a few Ringed Plovers are all that are noteworthy.

21st March.—Southerly breeze; clear.

The additions since yesterday are the first Wheatear of the season, a male; a Jackdaw; and an increase in the numbers of Rooks. A female Stonechat, a Mallard, and two Golden Plovers were also seen.

22nd March.—Only the dregs of the big flock of Lapwings which arrived on the 7th are now left; some thirteen seen to-day.

23rd March.—Strong south wind.

Lapwings, about fifty new arrivals to-day. Many immigrant Starlings in company with Lapwings.

Mealy Redpoll, one. Dunlin, one. Wood-Pigeon, one.

24th March.—South, strong breeze; clear.

A Scaup and a Woodcock are fresh arrivals. Two Herons, two Mallards, two Pied Wagtails, and many Blackbirds in since yesterday. In the early morning many immigrant Starlings struck the lantern of the South Lighthouse. (Weather hazy.) These arrived in considerable numbers, for they far outnumber the native birds.

25th March.—South, strong breeze; clear; rain last night.

Many fresh immigrants are in evidence to-day. The first Goldcrest, Lapp Bunting (male), four Swans, several Mallards, a Short-eared Owl, several Wigeons, one Woodcock, and an increase in the numbers of Lapwings and Blackbirds noted. A Wood-Pigeon, a male Stonechat, and three Pied Wagtails were also seen. The Peregrines were busy among the migrants, and one was seen to catch a Starling, while another was in pursuit of a Woodcock.

26th March.—Continued south wind.

The first Reed-Bunting (male), a Yellow Bunting, and an adult Black-headed Gull are new arrivals. Three Herons, three Woodcocks, and an increase in the numbers of Lapwings also noted. A Coot, seen this morning, is new to the Fair Isle list.

27th March.—Wind south, but too strong for birds.

Yellow Bunting, Jackdaw, and Rooks again seen to-day.

28th March.—Teal (male), Coot, Pied Wagtails, and Yellow Bunting still here.

29th March.—South-west; showers.

A Linnet, a Brambling, and about two hundred Skylarks seen.

30th March.—South-west; clear.

Many of the birds recorded still present—Rooks, a Jackdaw, Black-headed Gull, Common Gulls, Mallard, and two Mergansers, male and female. Decrease in the numbers of Lapwings.

31st March.—North-west, strong; clear.

Two Linnets noted; and a decrease in the number of Skylarks, Starlings, and Lapwings evident.

3rd April.—Bands of Puffins here for first time. A Brambling as an arrival.

5th April.—Many Chaffinches have arrived.

6th April.—To-day the Chaffinches are gone.

7th April.—South-west; light showers.

A few more immigrant Rooks to-day. Several Wigeon and Mallard again seen. Lapwings decreased to three.

9th April.—Several Snipe arrived since yesterday.

10th April.—South, light, fine; clear.

An adult male Lapland Bunting observed and heard singing. A Hedge-Accentor and two Jackdaws are new arrivals. Blackbird heard singing.

11th April.—South.

A few immigrant Bramblings in to-day, also two Grey Wagtails.

13th April.—South, light.

A Hedge-Accentor, a Redbreast, and a Linnet observed. Rooks and Jackdaws still here, but no Lapwings seen.

14th April.—East-south-east, light breeze; clear.

Three Wheatears (two males and one female), several Bramblings and Chaffinches, three Lapwings, and six

Golden Plovers—all arrivals. Puffins very numerous at breeding-grounds on the cliffs.

15*th April.*—Easterly breeze; clear.

More arrivals to-day. The first Whimbrels (a few), a White Wagtail, and several more Wheatears (ten males and a few females), seven Golden Plovers, and four Lapwings.

16*th April.*—North, light; clear.

A Yellow Bunting and a Mealy Redpoll noted.

17*th April.*—North-west wind.

Nothing seen to-day; birds are scarce.

18*th April.*—North-east, strong breeze.

One Grey Wagtail, two Pied Wagtails, and a White Wagtail seen.

24*th April.*—Wind east.

Short-eared Owl and Yellow Bunting as immigrants. Corn-Bunting and two Chaffinches seen, probably also arrivals.

25*th April.*—Wind changed to north-east last night, very strong, with thunder and lightning; changed to east, strong breeze, this morning.

Five Rooks, three White Wagtails, one Song-Thrush, and about twenty Common Gulls (young and old) observed.

26*th April.*—Easterly, light breeze; clear.

Numbers of immigrants present to-day. Many Wheatears have arrived, and are over all parts of the island. They are generally to be seen in scattered parties of twenty or so, and among the numbers present only two females were seen. Three Song-Thrushes, a Redwing, a Ring-Ouzel, and a Great Grey Shrike are also arrivals. A Fieldfare, two Hedge-Accentors, two Wood-Pigeons, a few Golden Plovers, and several Lapwings are also present.

27th April.—Light easterly wind.

Of the numbers of Wheatears seen to-day (over one hundred), only three were females. A flock of forty Bramblings has appeared, also two Willow-Warblers, three Kestrels, and numbers of Redbreasts. About twenty Song-Thrushes, a dozen Ring-Ouzels, about fifty Hedge-Accentors, and numbers of Fieldfares have arrived. White Wagtails are also numerous.

28th April.—South-east breeze; clear.

A few more migrants seen to-day. Among these, an adult male Tufted Duck is a new bird to the fauna. A male Blackcap has also appeared. Most of the migratory species seen yesterday have passed on—only six Ring-Ouzels, four or five Redbreasts, one Willow-Warbler, and the Hedge-Accentors are very much scarcer. The Wheatears are also decreasing, but the White Wagtails and Song-Thrushes are still plentiful.

29th April.—East, light; clear.

Redbreasts have again increased in numbers, about fifty birds seen to-day. About a dozen Chaffinches have arrived, mostly females. Four Willow-Wrens, a Mealy Redpoll, ten Fieldfares, and a few Ring-Ouzels also came under notice. The Hedge-Accentors and the Wheatears have decreased in numbers.

30th April.—East-south-east light wind this morning, and migrants plentiful.

A Lapp Bunting, a male, in nearly full summer plumage, was observed and heard singing. Thirteen Whimbrels, a Short-eared Owl, a Corn-Crake, several Swallows, and a Golden Plover, showing white on the primaries, have arrived since yesterday. Ring-Ouzels and Bramblings are numerous. Chaffinches have increased in number, and Wheatears are again plentiful and include several females.

1st May.—East-south-east breeze; very dull and wet, and the birds skulking and difficult to find.

Since yesterday two pairs of Wigeon, two Teal, and several Willow-Wrens have appeared. A few Hedge-Accentors and several White Wagtails were also seen.

2nd May.—Easterly light breeze; clear.

More female Wheatears to-day, many of the male birds have passed on. Three Lapwings and one Water-hen are arrivals. The species mentioned yesterday again seen to-day, and a Lapp Bunting and a Wood-cock also came under observation.

4th May.—With the light, south-easterly breeze many Wheatears of the large race (*S. leucorrhoa*), have appeared, and are to be found in all parts of the Isle. An increase in the numbers of Hedge-Accentors also noted. Eighteen Fieldfares seen; but the Song-Thrushes and White Wagtails are decreasing. A Common Wheatear observed carrying nesting material in its bill.

5th May.—This has been one of the best days of the season for migrants. The morning was uneventful, but towards evening numbers of migrants were observed all over the Isle. Light rain fell in the morning, with the wind in the south-east; but later in the day the weather cleared and the wind backed to the east.

The first Common Sandpiper for the season, numbers of Chaffinches (mostly females), and several Snipe were observed. Two Great Snipe were put out of some rough grass. These were the chief arrivals seen up to noon.

Afterwards, two Mealy Redpolls (*A. linaria*), a female Grey-headed Wagtail (*M. thunbergi*), a Tree-Pipit, Willow-Warblers, a Short-eared Owl, the first Reed-Buntings, two Swallows (one of which was captured in a house), and a Corn-Crake were seen. In the morning only about twenty Fieldfares came under notice, but about 4 P.M. they were everywhere, and vast flocks could be seen circling overhead. In company with these thousands of Fieldfares were many Ring-Ouzels and Starlings. A flock of one hundred Bramblings was seen, and many large Wheatears (*S. leucorrhoa*) were also present. Rain again began to fall about 3 P.M., and this no doubt spoiled what would otherwise have been a most eventful day among the migrants, for the dull wet weather drove the smaller birds to seek shelter in the ditches, cliffs, and walls, in fact, any hiding-place that could be found. Warblers were probably far more numerous than they appeared to be.

6th May.—Yesterday evening the south-east wind freshened to a strong breeze, bringing fog with it, which, however, cleared off this morning about 8 A.M. Weather dull and hazy.

The newcomers noted are an adult male Subalpine Warbler, Mistle-Thrush, Redstart, Common Whitethroat, Green Sandpiper, male Pied Flycatcher, several Mealy Redpolls, Tree-Pipit, Long-eared Owl, Common Sandpiper, three Black-headed Gulls (adults), male Blackcap, and a large increase in the numbers of Fieldfares. The vast concourse of the latter species now present, prevents any estimate being formed of their numbers, but there are probably not less than three thousand, and they are to be seen in every part of the Isle. The Mistle-Thrush mentioned was in their company, and it is possible more of these birds were present. Redbreasts numerous, but skulking in the ruins of the old walls. Several Dunlins also seen.

7th May.—Wind changed from south-east to west, with heavy rain.

The first Whinchat and Sedge-Warbler of the season appeared. More Black-headed Gulls, several House-Martins, more Pied Flycatchers, another Grey-headed Wagtail (*M. thunbergi*), several Common Sandpipers, several Woodcocks, a few Tree-Pipits, and another Reed-Bunting have arrived. A Merlin, several Ring-Ouzels, a Heron, a drake Mallard, a Teal, several Willow-Wrens, many Swallows, and a number of Dunlins were also observed through the day. Several of the large race of Wheatear (*S. leucorrhoa*) were present, and a flock of forty Lapwings completes the list of the species seen.

8th May.—The wind having changed from the south-east to west, a less favourable quarter, birds are much scarcer. In the evening the wind backed to southerly: a light breeze; clear.

Two Corn-Buntings and a Corn-Crake were noted as arrivals. A Redwing and several Swallows came under notice. The flocks of Fieldfares have greatly decreased, leaving only a scattered remainder. The Redbreasts and Hedge-Accentors have also passed on.

9th May.—Wind south-east, with rain.

Over two hundred Fieldfares have arrived since last night, also a few Redbreasts and Hedge-Accentors. Two Grey-headed Wagtails, an Ortolan Bunting, several Snow-Buntings, a Garden Warbler, several Whinchats,

Pied Flycatchers, several Tree-Pipits, and a Corn-Crake have also appeared.

10*th May.*—East-south-east, strong breeze and rain all day; very little seen, all the migrants in hiding.

11*th May.*—Fine south-east breeze; clear.

A Golden Oriole (a female) was found dead to-day in the crevice of a cliff, where it had crept for shelter. This is an addition to the Fair Isle list. Another new bird for the season, a Yellow Wagtail (a male), has also appeared. The other birds seen to-day were seven Mealy Redpolls, Willow-Warbler, Common Whitethroat, Whinchat, Sedge-Warbler, Little Grebe, and three large Wheatears (two males and a female). The Hedge-Accentors have gone, and only two Redbreasts are left. A single Ring-Ouzel, a flock of two hundred Fieldfares, and a male Snow-Bunting in full summer plumage, also came under notice.

12*th May.*—The good weather with the favourable south-east wind continues, and many arrivals have again been noted.

Four Grey-headed Wagtails (one a beautiful adult male) were seen, and a Little Bunting was observed at close quarters, after which it joined a horde of Twites and was lost. An Ortolan Bunting, a Tufted Duck, a Siskin (male), three Redstarts (females), more Tree-Pipits, several Common Whitethroats, a Short-eared Owl, and two Swallows were the other birds noted as arrivals. The Common Wheatears have also greatly increased in numbers, and several Willow-Warblers were seen. The Fieldfares have decreased, and the Song-Thrushes, Redwings, Ring-Ouzels, Hedge-Accentors, and larger Wheatears have passed on. The Redbreasts are also scarce, only two or three being seen.

13*th May.*—Light easterly breeze and fine weather.

A Spotted Flycatcher, the first of the season, has put in an appearance; and the other arrivals include Common Whitethroats, Willow-Warblers, a few more Rooks and Swallows, and about a dozen Whinchats. Two Jackdaws, a Pied Flycatcher (male), one Ring-Ouzel, and three House-Martins complete the list of migrants for the day.

14*th May.*—This was one of the red-letter days of the season. The weather was fine and settled.

Two additions were made to the avifauna of the island, namely, Savi's Warbler (female) and a Hawfinch (male). The Savi's Warblers (there were two of them) were the wildest birds I (George Stout) ever came across. They were pursued up and down a burn for hours ere one of them was obtained for identification. When disturbed, they would rise from among the cover and fly for a short distance, and then literally dive into the grass, etc., with which the sides of the ditch were thickly clothed. One of them was actually observed swimming across a small pool of water. A Grasshopper Warbler and several Sedge-Warblers were seen at the same place. The Hawfinch was found on the northern part of the Isle, and when first seen, was digging its bill into pony's dung, and was so engrossed with its occupation that it allowed a close approach. The other arrivals noted were as follows: the first male Redstart, an Ortolan Bunting (male), a Pied Flycatcher (male), five Red-breasted Mergansers, three Common Gulls (which passed on at once), nearly twenty Dunlins, and numbers of Tree-Pipits.

15*th May.*—With the moderate south-east breeze more birds have arrived.

The first Cuckoo for the season, also the first Lesser Whitethroat were observed. The other new arrivals were a Chiffchaff, a Wood-Pigeon, several Redstarts, and another Ortolan Bunting. A Garden-Warbler, a Willow-Warbler, and several Tree-Pipits were again seen. Many Whinchats have passed on, for only three were noted to-day.

16th May.—South-east, moderate; fine, sunny. The weather conditions being still favourable, birds are again plentiful.

Several more Redstarts (males and females) have arrived, also Sedge-Warblers, Blackcaps, Greater Wheatears, and another Cuckoo. The Willow-Warblers, Chiffchaffs, and Whinchats are more numerous, and an increase is also noted in the numbers of Redbreasts, Swallows, and House-Martins. A Garden-Warbler, a Spotted Flycatcher, and a Grey-headed Wagtail may also be new arrivals. Eleven Tree-Pipits were seen together; there are about forty present. Common Sandpipers were also noted.

18th May.—Wind has changed to bad quarter, north-west.

The birds noted were probably here yesterday. These were four Blue-headed Wagtails, a Spotted Flycatcher, a Reed-Bunting, several Willow-Warblers, numerous Common Whitethroats, two Whinchats, two White Wagtails, two Chiffchaffs, and a Garden-Warbler. The Tree-Pipits are still very numerous, but the only members of the Thrush family present are two Blackbirds.

19th May.—Westerly breeze; birds scarce.

A Linnet, a Blackcap, an Ortolan, and a Whinchat were observed. The Tree-Pipits are decreasing, and only a few Willow-Warblers are now to be seen.

20th May.—Light westerly breeze.

The first Jack Snipe of the season was observed, also a Hedge-Accentor and Greater Wheatear.

21st May.—South-west, light breeze, morning; changed to south, with fog, in the evening.

The first Sand-Martin appeared, also several Swallows and House-Martins. Tree-Pipits still decreasing.

22nd May.—The first Red-backed Shrike and another Sand-Martin have arrived. Only one Dunlin now left, and the Tree-Pipits are gradually passing on. Two White Wagtails were again observed.

23rd May.—Southerly breeze; clear.

Two Manx Shearwaters seen off the island. The birds noted on land were six Yellow Buntings, a Greater Wheatear (female), a Teal, two Sedge-Warblers, and a Fieldfare.

25th May.—Southerly breeze, morning; changed to west, evening.

Common Whitethroats, two; Garden-Warbler, one; Sedge-Warbler, one.

26th May.—West, light breeze; clear in morning; changed to south in afternoon.

A Wood-Sandpiper, the first for Fair Isle, was disturbed while feeding on the side of one of the mill-dams. It was extremely wary. A Sedge-Warbler, several Greater Wheatears, two Common Whitethroats, a Garden-Warbler, two Grey-headed Wagtails, a Common Gull, several Black-headed Gulls, a Spotted Flycatcher, and a male Wigeon were the other birds noted. Over twenty Swallows and several House-

Martins were observed capturing the numerous flies that were hatching out of the decaying seaweed at high-water mark.

27th May.—South-east breeze, with haze.

The birds to-day are: two Turtle-Doves, two Common Whitethroats, and a male Redstart.

28th May.—Strong westerly breeze. Nothing worth recording.

29th May.—South-west breeze; hazy.

The following birds seen: male Blackbird, Sedge-Warbler, Garden-Warbler, Willow-Warbler, Spotted Flycatcher, and Turnstone.

30th May.—Wind south-east, with fog.

The birds in to-day are: a Lesser Whitethroat, a Short-eared Owl, a Cuckoo, a Common Whitethroat, three Golden Plovers, and several Whimbrels.

31st May.—With the light easterly wind overnight, many birds have arrived.

An adult male Red-spotted Bluethroat was seen to good advantage, likewise a Black Redstart. The other arrivals are: Blackcaps, two females; Sedge-Warblers, two; several Lesser Whitethroats; two Scaups, male and female; and two Redshanks. In the evening a small stranger was observed at fairly close quarters through the field-glasses, and there can be no doubt that it was an adult male Red-breasted Flycatcher. This bird would occasionally erect its tail almost, if not quite, above the level of its head; and its other actions greatly resembled those of the Spotted Flycatcher.

1st June.—The easterly wind continues, and despite a raw, damp fog, many immigrants are in evidence.

About a dozen Red-spotted Bluethroats were seen in different parts of the island. These birds, however, are so very shy that many were doubtless missed, and the fastnesses of cliffs would also afford shelter for a number of them on such a day. When pursued, these birds ran at incredible speed along the bottoms of the dry ditches which they frequented, and it was extremely difficult to get near them. Another new and interesting visitor was an Icterine Warbler. The other arrivals were: several Red-backed Shrikes; Common Sandpiper, one; Blackcaps, three males and two females; a Short-eared Owl; Spotted Flycatchers, several; Garden-Warblers, three; Turnstones, two; Lapwings, four; Whimbrels, several; Willow-Warblers, two; Whinchats, five; Greater Wheatears, several; Lesser Whitethroats, numerous; Common Whitethroats, several; Grey-headed Wagtails, three or four; Tree-Pipits, several; Sand-Martins, two; House-Martins and Swallows, many; Chiffchaffs, two; several female Redstarts; Teal, male; and Dunlin, one. No doubt the birds were far more numerous than these records indicate, for only those which actually came under notice are mentioned; other species, too, would escape notice.

2nd June.—Wind still easterly in morning; changed to south-east in evening.

The dregs of the rush remain. Two Bluethroats were seen; likewise a Redstart (adult male), several Blackcaps (females), a Corn-Bunting, and three Garden-Warblers.

3rd June.—The migrants have passed on, the conditions being favourable, for the south-east wind still continues.

A male Whinchat, a Common Whitethroat, two female Redstarts, two Willow-Warblers, and three

Blackcaps (two females and a male) are all that were seen.

4th June.—Westerly light breeze; rain.

A Wryneck was observed on the face of the cliffs at the South Harbour, and a male Red-backed Shrike was also seen.

5th June.—Northerly breeze; clear.

A female Blackbird and four Black-headed Gulls were all the visitors seen to-day. A Common Sandpiper and several Purple Sandpipers left about this date.

6th June.—North-east breeze; clear.

The birds seen to-day were: Tree-Pipit, one; Common Whitethroat, one; Redstarts, two females; Corn-Bunting, one; Cuckoo, one. Several Swallows and House-Martins are still on the Isle.

7th June.—The last Snow-Bunting seen.

8th June.—East, light wind; clear.

Lesser Whitethroats, four; Garden-Warbler, one; Chaffinches, two females; Tree-Pipit, one; Whimbrels, twenty-four; Common Sandpiper, one; Blackcaps, male and female; House-Martins, several; and three Chimney Swallows. Some of the birds mentioned are fresh arrivals.

9th June.—South, light wind; clear.

Common Gulls, two; Black-headed Gulls, two; and a Dunlin, are all that were seen.

10th June.—South, light wind; clear.

A Garden-Warbler and a Sedge-Warbler came under notice. The Swallows and House-Martins are still here.

11th June.—South-west breeze; clear.

A Sanderling seen. The Common and Black-headed Gulls still here.

12th June.—Blackbird seen, and heard singing. Three Wigeon, a male and two females, observed.

13th June.—A pair of Lapwings still here.

15th June.—South breeze; clear.

A Kestrel and a male Blackcap observed to-day. An albino Puffin seen among a crowd of normally coloured individuals.

16th June.—Southerly strong breeze; clear.

A Two-barred Crossbill (*L. bifasciata*), an adult male, was observed. Like the Hawfinch, it had to resort to seeking food among the dung of the ponies. A Wood Pigeon was seen, and several House-Martins are still present.

17th June.—Westerly light wind; clear.

Five Lapwings arrived to-day, and at once passed on.

18th June.—Westerly light wind, early; changed to south-east later in the day.

Two interesting visitors were detected to-day—namely, a female Short-toed Lark, and an adult female Rustic Bunting. The latter was disturbed from among the long grass in the peat banks, and as it flitted, uttered a short note very like that of the Reed-Bunting.

20th June.—Northerly strong breeze; clear.

A Turtle-Dove and a Lesser Whitethroat seen.

21st June.—Two Whimbrels still here.

22nd June.—The following birds still here:—Golden Plover, one; Chaffinch, one; Turtle-Dove, one; Swallow, one; and House-Martins, several.

23rd June.—The first four Swifts seen this year appeared; three Black-headed Gulls are still here.

27th June.—Turtle-Dove still here. A White Wagtail observed.

28th June.—One Heron seen.

29th June.—Three Curlews and two Whimbrels came under notice.

30th June.—Four Swifts, a Golden Plover, and two Swallows observed. The occurrence of the Golden Plover probably indicates the beginning of the movements that take place at the close of the breeding season.

1st July.—Corn-Bunting and House-Martin seen.

2nd July.—A few Swallows have arrived. The Heron again seen.

5th July.—Golden Plover seen.

6th July.—Two Arctic Skuas observed in the South Harbour. A Tern, a Swift, and several Black-headed Gulls observed.

7th July.—Three Whimbrels present, and a Manx Shearwater seen off the island.

8th July.—Eight Lapwings have appeared on the isle, more Black-headed Gulls are present, and five Herons were observed.

9th July.—About twenty Black-headed Gulls present; also five Common Gulls. The Lapwings have increased to twenty.

11th July.—Song-Thrush seen. Several Curlews and Whimbrels present. Lapwings decreasing.

13th July.—Several Redshanks have appeared. Two Herons still here.

14th July.—Four immature Black-headed Gulls present.

15th July.—One Swift seen.

16th July.—Three Golden Plovers.

20th July.—Many Redshanks have appeared all round the island.

22nd July.—Pied Wagtail seen. Heron and Black-headed Gulls still here.

28th July.—Two Dunlins observed. Redshanks plentiful.

30th July.—Swift seen.

31st July.—Two Pied Wagtails appeared.

1st August.—Several Black-headed Gulls, new arrivals, and all immature. Swift, one; Wigeon, three; Heron, one.

2nd August.—A Dunlin and a Mallard observed.

3rd August.—Terns (two) seen; also two Common Gulls.

4th August.—A few Pied Wagtails here to-day; also four Terns and ten Lapwings.

5th August.—Nine Curlews noted.

8th August.—Two Golden Plovers, several Pied Wagtails, a Mallard, Common Gulls (several young and old), Black-headed Gulls (many young and a few adults), and twenty Curlews here (several of which are new arrivals).

10th August.—Sanderling and Mallard observed.

11th August.—First White Wagtail for autumn to-day; also a Swift and three Golden Plovers.

13th August.—North breeze; clear.

An immature Temminck's Stint to-day. It frequented a small muddy pool in the north of the island. It is an addition to the avifauna. A Knot seen in company with three Golden Plovers. About forty Common Gulls and many Black-headed Gulls present.

14th August.—Three White Wagtails and two Turnstones are arrivals.

17th August.—North-east, light; clear.

A Song-Thrush, a Tern, and seven Turnstones seen. About thirty White Wagtails (mostly young birds) arrived this evening.

18th August.—Swift and Knot observed.

19th August. — Two Greater Wheatears (*S. leucorrhoa*) observed.

20th August.—About seventeen Herons as new arrivals.

21st August.—East-south-east; clear.

Sanderlings, Dunlins, Turnstones (nearly twenty), Curlew (about twenty), Knot, Lapwing, Greater Wheatear, and about fifty White Wagtails (mostly young birds). Remains of Nightjar found.

22nd August.—Easterly, light; clear.

In early morning a flock of Waders was observed, comprising about forty Redshanks, six Curlews, one Whimbrel, and a few Dunlins. Other birds observed were:—Terns, three; Sanderlings, two; Knot, one; Greenshank, one observed on reefs; Greater Wheatear, and Common Snipe.

24th August.—East, light; wet.

Greater Wheatear, two Whimbrels, Snipe, Reeve, and four Willow-Warblers. Knot and Greenshank again observed.

25th August.—East-south-east breeze; rain.

First Lesser Whitethroat for autumn, a few more Turnstones, and two Lapwings seen. Greenshank left about this date.

26th August.—Easterly, light; sunny.

Another Knot, a Wood-Warbler, a Tree-Pipit, several Dunlins, two Snipe, about a dozen Willow-Warblers, and two Sanderlings, are new arrivals. Swift and Mallard also seen, and White Wagtails are plentiful.

27th August.—Very strong westerly breeze.

No new birds in. The following observed :—Turnstones, Knots, Lapwings, Golden Plovers, Willow-Wrens.

28th August.—South-west ; clear.

Sanderlings, three ; Song-Thrush, one. Many Redshanks and the two Knots left about this date.

29th August.—West breeze ; clear.

Cormorants, many arrivals for the winter.

31st August.—North-west breeze ; clear.

Ten Lapwings, four Sanderlings, and one Tree-Pipit. Cormorants in numbers seen flying over isle.

1st September.—One Whinchat.

2nd September.—North-west breeze ; clear.

Herons, four ; Teal, one ; Golden Plover, and a Lesser Whitethroat.

3rd September. — Great Skua seen passing over isle.

4th September.—Wind north-east, chilly ; very cold at night.

The author landed at Fair Isle in the morning, and in the afternoon had a round of the crofts with George Stout. Visited nests with chicks of the Storm Petrel. They were rugged nurseries amid a chaos of stones. One of the chicks was only a few hours out of the shell, and the mother was in attendance. In the other case the babe was a few days old, and the parents were absent, as in other instances that have come under my notice. An adult female Scarlet Grosbeak, observed on a fence adjoining a field of bere (barley). Its stomach contained grains of bere and a few white stones. House-Martin, one on face of cliff, hawking flies.

5th September.—Fine, but dull, A.M. South-west gale and rain, P.M. Wind moderate; fair later.

Barred Warbler, in first plumage, and the first of the species obtained on Fair Isle. Heron, seven seen in party; Whimbrel, one heard; Sanderlings, a few; Dunlin, one; Turnstones, a few; Purple Sandpiper, first of season; Redshanks, several; Curlews, several.

6th September.—West, moderate, cool; showery.

Wheatears, a number of the large race; Kestrels, several; Sanderlings, three; White Wagtail, one.

7th September.—Light, west; sunny, warm. South-east, light in afternoon and at night.

Barred Warbler, an adult male on the face of the cliff: it was very wild. Lapwing, an immigrant seen; Kestrels, several seen; Merlins, two seen; House-Martin, one; Sanderlings, two; Song-Thrush, an immigrant; Ring Plovers, several immigrants.

8th September.—A poor day for bird-observing, the high wind and rain causing the birds to seek and remain in shelter.

Song-Thrush, an arrival; Turnstones, several arrivals.

9th September.—North-east gale with heavy rain until 3.30 P.M. Saw a Bar-tailed Godwit, the first I have observed here. It was exceedingly tame, and was walking on the grass. All other birds skulking, owing to the high wind and heavy rain. The day grew suddenly fine at 3.30 P.M., and the birds which had arrived overnight then showed themselves.

Mealy Redpolls were evidently present in some numbers, as several were seen in different parts of the island. Redstart, one seen, an arrival; Willow-Warbler, several seen; White Wagtails, more numerous; Tree-Pipit, one; Lapp Bunting, one; Snow-Bunting, one; and Temminck's Grasshopper Warbler (*L. lanceolata*), a young bird: the latter a new visitor to the island.

10*th September.*—North, strong and blustery, sunless, cold. All birds sheltering.

A Green Sandpiper, a Mealy Redpoll, a Tree-Pipit, a Blackcap, and two Ortolan Buntings seen.

11*th September.*—North, moderate; showers early, later sunny, but cool.

A young male Red-backed Shrike, several Mealy Redpolls, two Ortolans, several Tree-Pipits, a Lapp Bunting, and a Barred Warbler seen; the last-named took refuge among oats. Lesser Black-backed Gulls, adult and young, seen daily to date. Ring Plover, fourteen immigrants in flock.

12*th September.*—North, moderate, cool; sunny and showery.

A Common Whitethroat among potatoes, several Ortolan Buntings, a Snow-Bunting, a Tree-Pipit, and two Lapp Buntings observed; only one White Wagtail. This species is now almost entirely absent from the island.

13*th September.*—South-east and southerly airs, fine at first; then south-west, light; showery.

A male Red-breasted Merganser in South Harbour. Barred Warbler in the cabbages. Ortolan on face of cliff, on west side.

14*th September.*—South-west, rain in morning; north-west, fine, sunny later.

An Ortolan Bunting, a Song-Thrush, and two Tree-Pipits. All other migrants scarce. Noticed that the

Twites are very partial to the heads of the thistles, and crowd upon them, paying very little attention to the observer.

15*th September.*—Light, variable airs; but brilliant morning. Dull, north-west in afternoon and at night.

Four Wigeon—the first seen of the season—a male, female, and two young birds. Two Merlins, two Kestrels, a young male Red-backed Shrike in very red-brown plumage, and a Barred Warbler among cabbages. All migrants scarce.

16*th September.*—South-east, light at first, increasing to strong breeze.

An adult male Lapp Bunting, several Greater Wheatears, and three immigrant Lapwings seen.

17*th September.*—South, rain till 9 A.M.; then south-west breeze, fine.

Six immature Wigeon in the North Haven; a number of Greater Wheatears have arrived since yesterday. A Whimbrel, a Dunlin, and two Kestrels still here; also a few Lesser Black-backed Gulls, but most have gone. Many young Herring Gulls that persist in hanging about the houses are slowly starving.

18*th September.*—South to south-west, rain in forenoon; then south to south-east, fine.

An adult Willow-Warbler, three Dunlins, several Golden Plovers, a young Turtle-Dove, a Snipe, several Whimbrels, many Greater Wheatears, and twelve Wigeon seen. A disappointing day. At night from 8-10 P.M., the rays from the lantern were conspicuous, and a party of about six Redshanks flew noisily around the lighthouse for a short time. A few Wheatears also approached the light, and a Storm Petrel was captured and liberated.

19*th September.*—South-south-east; rain all day. Very few birds observed.

A young Willow-Warbler, a Turtle-Dove, a few Golden Plovers, and a fair number of Greater Wheatears seen.

20*th September.*—South-south-east, moderate. Fog from 2.30 A.M. until 3 P.M. High wind and heavy rain at night.

A Common Whitethroat, a Lapp Bunting, and a Red-backed Shrike seen. The latter was seen to pursue a Wheatear and to stoop at it several times.

21*st September.*—South-east, light. Fog from 10 A.M. to 4 P.M.

Garden-Warbler and Jack Snipe killed at lantern at 3.40 A.M., and several Wheatears flying in the rays. The weather was hazy, and the beams of light very conspicuous. Several Willow-Warblers, a young Swallow, a Golden Oriole, a Snipe, and two Water-Rails seen among potatoes.

22*nd September.*—South-east, moderate breeze; dull. A considerable arrival of migrants over night. The first marked immigration of the month.

A Willow-Warbler and Meadow-Pipit killed at lantern at 12.30 A.M. Skylarks observed emigrating towards the Orkneys, at 6.30 A.M. A Lesser Whitethroat, several Siskins, a Song-Thrush, many Redstarts, several Common Whitethroats, a male Whinchat, many Willow-Warblers, an immature Rustic Bunting, many Tree-Pipits, a male Blackcap, four Kestrels, a female Scaup-Duck, one Mallard, and a large flock of Golden Plovers noted. A Pied Flycatcher at the lantern at 8 P.M.

23*rd September.*—South-east, light breeze. Fine day,

with dull and light periods. The best bird day for autumn as yet.

A number of Red-spotted Bluethroats (adults and young). These birds were, with one exception, found in potatoes or turnips. They ran with the rapidity of a Partridge along the drills, sometimes for a hundred yards, until they reached the end of the cover; then with a rapid flight they doubled back and repeated their tactics. They are dark like a Hedge-Accentor on the wing, and the bay-red tail with its broad dark terminal band is not so conspicuous as one would suppose. One of the birds was found by the side of a small stream, a haunt much resorted to, where there was little or no cover; and it ran under a stone for concealment, which is also a common habit. Three Ortolan Buntings were seen. These birds affect the corn, and one was feeding on the stooks. A Reed-Warbler was raised in the potatoes, and sought shelter in standing oats, where it was shot to establish its identity. Whinchat, yesterday's bird, seen. Blackcaps, male and female, and Garden-Warbler in potatoes. Willow-Warblers, Tree-Pipits, Snipe, and Redstarts numerous. Chaffinch (first of season), Ring-Ouzel (first). White Wagtails, a few (the first since the 12th); Red-breasted Mergansers, three on sea; Golden Plovers, many. Grey Plover, one with Golden Plover; Kestrels, several seen. Common Sandpiper flying in rays for some time at 6.30 P.M. Jack Snipe killed at lantern during night.

24th September.—South-east, light breeze; dull, and mild.

Many birds have arrived since yesterday. Redstarts particularly abundant. Pied Flycatcher killed at lantern during the past night, and a number seen on cliff-face and walls. Bramblings, a few only—the first. Lesser Whitethroats, several. Chaffinches in fair numbers. Water-Rail, one by ditch side. Siskins, a few. Whinchats, two seen on stooks. Redwings, a fair number in the crofts, and many on the north-west cliffs—the first of the season. Willow-Warblers, Garden-Warblers, and Blackcaps (both sexes), numerous. Reed-Bunting, two seen. Redstart (both sexes), very common. White Wagtail, one seen. Tree-Pipit, fairly numerous. Swallow, an adult. Goldcrests, a number on the cliffs. Spotted Flycatcher, one. Ortolan Buntings, three seen. Lapp Bunting, one. Kestrel and Lesser Black-backed Gull still here. Whimbrel, one. Common Sandpiper heard overhead at 7 P.M. Great numbers of birds on the face of western cliffs (where insects were abundant), but beyond the range of identification.

25th September.—South-east, light breeze; bright sunshine—a beautiful day. Birds very numerous, especially on the face of the great cliffs.

Lesser Whitethroats, several seen. Grey Plover, one. Swallows, a number. Chaffinches, many on the stooks and on stubbles. Bluethroats, three seen—two among potatoes and turnips, and one in garden, perched on a clothes line. Willow-Warblers everywhere, many on face of the cliffs. Common Whitethroats, two seen in turnips. Jack Snipe, one in potatoes. Blackcaps and Garden-Warblers, a number in the cover afforded by the crofts. Siskins, many in small parties of two or three, or singly. Redstarts everywhere, many on the cliffs. Tree-Pipits numerous, several on face of cliffs. Yellow-browed Warbler, one on face of north-west cliffs. Goldcrests, many on face of cliffs. Snipe, several. Little Buntings, two on stooks. Pied Flycatchers, a number on the cliffs. Redwings, a few. Reed-Bunting, three. Song-Thrushes, several in geos. Ring-Ouzel, one. Fulmar Petrel, one seen off the north-west cliffs. Common Sandpiper, one on rocks by the sea. Greater Wheatears, most numerous on high ground, but in fair numbers elsewhere. Skylarks, large additions to numbers, all over stubbles and grasslands. Jack Snipe killed at lantern last night.

26th September.—South-east, light; dull, warm, sunless. Fog-horn at 11.15 P.M. Weather still in favour of passage, and many birds in to-day.

Wheatear and Jack Snipe killed at lantern at 2 A.M. Chiffchaffs in potatoes and cabbages. Bluethroats, two in cabbages. Blackcaps, a few. Lesser Whitethroat, one seen. Ortolan, three seen. Siskins, many, twelve in a party; they are noisy little birds, and always proclaim themselves. Reed-Warbler in potatoes; this bird always looks very red when on the wing. Reed-Buntings, several. Goldcrests on face of cliffs. Pied Flycatcher on face of cliffs. Song-Thrushes and Redwings, numerous in crofts. Bramblings, a few. Grey-headed Wagtails, a male and female. Chaffinches, many of both sexes. Kestrels, several. Lapp Buntings, two seen. Dunlin, several. Redbreast, one in cabbages. Sand-Martin, one. Jack Snipe, two in crofts. Willow-Warblers, common in the crofts, and many on the cliffs. Spotted Flycatcher, one. Swallows, several. Herons, six in party. Greater Wheatear, many. Tree-Pipits, many, a considerable increase. Whinchat, one. Redstarts, a number. Little Bunting killed at north lantern between 12 and 3 A.M. Others said to have been present; crop quite empty, as is the case with nearly all the birds killed at the lanterns.

27th September.—Foghorn going until 2.30 P.M. South-east, light air; fine.

Many arrivals since yesterday. Yellow-browed Warblers, three seen; one, on rocks near lighthouse, very tame, and allowed close approach; flicked its wings like a Willow-Warbler; the other two were on the face of the cliffs. Redstarts in fair numbers on cliffs. Greater Wheatears common. Song-Thrushes and Redwings plentiful on cliffs. Merlin on moor. Pied Flycatcher on cliff. Sanderling on beach. Tree-Pipit on cliff. Jack Snipe by burn side. Little Stint by burn; very tame. Chaffinches, a flock of fifty, composed of both sexes. Chiffchaff, amid dockens. Eider Ducks, an immense flock, nearly all males in various stages of pied plumage, but majority in full dress. Willow-Warblers on cliff. Reed-Bunting, one.

28th September.—South-east, breezy; dull, cooler; wind more southerly at night. Foghorn at 7.30 P.M.

Starlings, first immigrants, at lantern at 3 A.M. Yellow-browed Warbler on face of cliff near lighthouse. Siskins, a few feeding on thistles. Swallows, about six seen in pairs at different localities. Bluethroats, three seen in potatoes, turnips, and brackens. Whinchat, one. Tree-Pipits, several. Redstarts, not so common, but still in numbers. Goldcrest, one on cliff. Redwings and Song-Thrushes on cliffs, and not so numerous. Bramblings, a few. Blackcaps, two in potatoes. *Phylloscopus borealis*, a male, a beautiful specimen; it was found in turnips, and was very wild its dark colour attracted attention. White Wagtails, a few. Garden-Warbler, in turnips. Redbreast, in brackens. Greater Wheatears, some. Snipe, several, one in potatoes. Kestrel seen. Water-Rail, one in rough grass, looked exactly like a Water Vole. Corn-Crake, one among corn.

29th September.—Foghorn off and on all night. South-south-east, light and fine. South in afternoon, breezy; some rain.

Willow-Warbler, one. Blackcap, female. Redstarts, not a few. Chiffchaffs, two in potatoes and dockens. Common Whitethroat in potatoes. Swallows, several. Lesser Whitethroats, several, among cabbages, turnips, and potatoes. Snipe, several, in turnips and potatoes. Little Buntings, two on stooks. Redbreast in garden. Reed-Bunting, one. Song-Thrushes, five in geo. Scarlet Grosbeak, young male among party of Twites in the stubbles. Jack Snipe, one only seen.

30th September.—South, light, dull; south-west, rain in afternoon; fine later.

Pied Wagtail, male adult; plumage a mixture of faded summer dress and fine black fresh feathers. White Wagtails, some. Garden-Warbler in potatoes. Reed-Bunting in turnips. Chiffchaffs, four found in potatoes and turnips. Redstarts, a few. Fieldfare, the first. Richard's Pipit, male found in a patch of potatoes, and alighted on wire fence; its note was loud and single; flight undulating, like that of Tree-Pipit, and not jerky like the Meadow-Pipit's. Tree-Pipits, a number. Yellow-browed Warbler in potatoes. Whinchats, two noted. Lesser Whitethroat in turnips. Kestrel again seen.

1st October.—South-east to west in A.M., light; showery at first, then sunny and hot; south-west, dull and oppressive in afternoon.

Birds few; most have moved on. Tree-Pipits, a few; nearly all gone. Redstarts decreasing. Song-

Thrushes and Redwings, a number on the face of the cliffs. Garden-Warbler, Lesser Whitethroats, and female Blackcap in crofts. Bluethroat, a young bird, came from cliff face and alighted in potatoes. White Wagtails, a few. Pied Wagtail, one. Grey-headed Wagtails, the two seen on the 26th ult. still here: they were very wary and frequented grassland, having a predilection for alighting close to the oxen, for the sake of the insects disturbed. Merlins, two seen; one, a male, made a stoop at a Peregrine, which it flew after for some time. Chiffchaff, one only seen. Chaffinches, very few now. Greater Wheatears, many; one at lantern at 3 A.M. Skylark at lantern at 12.30 A.M. Lesser Whitethroat at lantern at 10.30 P.M. Meadow-Pipit at lantern at 10 P.M. Greater Wheatears at lantern at 11.30 P.M. Turtle-Dove, immature, captured at north lantern at 11.30 P.M. Kestrels, two still on isle. Gannet fishing at 6.15 P.M., dusk.

2nd October.—South-east, light, misty, then sunny and hot; south-west, very light, P.M.

Starlings, immigrants, at the lantern from 1 to 3 A.M. Greater Wheatears at lantern from 11.10 to 3.15 A.M.; many on island. Red-throated Pipit seen and heard. Redstarts, very few. Hedge-Accentor (first), Willow-Warbler, Chiffchaff, and Whinchat seen; Tree-Pipits, fairly numerous. Bramblings, several. Chaffinches, a few. Jack Snipe in potatoes. Kestrels, two. Ringed Plovers, several. Teal, two.

3rd October.—West breeze, brilliantly fine; south-west later.

Very few birds as immigrants to-day; only the dregs of a few species. No Warblers in crofts. Bramblings, some in stubbles. Jack Snipe, two. Redstarts, two.

Greater Wheatears still common. Kestrel, one. Red-throated Pipit, an immature female: it rose from potatoes, and was detected by its strange note heard by me for first time in Fair Isle yesterday. Ring Plovers, several in flock. Merlin, one. Tree-Pipits, very few now. Hedge-Accentor, one among cabbages.

4th October.—North-west breeze, much cooler; dull, and then sunny periods; fog at night.

Redwing killed at lantern at 1 A.M.; Jack Snipe at 2 A.M. Redwings and Snow-Buntings, a number of arrivals. No Lesser Black-backed Gulls in their favourite haunts.

5th October.—South-east, dull, cool, A.M.; south-south-east breeze, P.M. Very few birds after yesterday's north-westerly weather.

Larks emigrating south at 8 A.M. Siskins, a small party. Jack Snipe in turnips. Bramblings, a flock of about fifty. Snipe in potatoes. Reed-Bunting, one. Goldcrest, male, in cabbages. Merganser in geo. Tree-Pipits, a few. Lesser Black-backed Gull, one immature. Merlins several. Wheatears, comparatively few now. Snow-Buntings, thirty in flock.

6th October.—Southerly breeze; dull, foggy.

Many arrivals since yesterday, but all in hiding and very difficult to find. Redwings and Bramblings at lanterns of both lighthouses at 4 A.M.; large flocks on the island; the former very numerous on the face of the cliffs. Jack Snipe, a number. Lesser White-throat in cabbages. Wheatears (both races), very few. Song-Thrushes, many on the cliffs and elsewhere. Tree-Pipit, one only. Alpine Accentor, one distinctly seen at close quarters on the face of the cliff, on the west side. Ring-Ouzel, one. Blackbird, one. Reed-Buntings, two. Siskins, a few. Meadow-Pipits, many.

7th October.—South, light; brilliant day.

Chaffinches at lantern at 3 A.M., and many of both sexes on isle. Bramblings at lantern at 3 A.M., and numerous in the stubbles. Blackcap in potatoes. Lesser Whitethroat in turnips. Jack Snipe, several. Whinchat, one in potatoes. Merlins, several sweeping over the stubbles in pursuit of small birds. Wood-Lark, one on the wing, uttering its pretty warbling note. Yellow-browed Warbler, one in garden. Meadow-Pipits, many.

8th October.—South-east, sunny and dull periods; increase of wind at night, with fog.

Blackcaps, two in cabbages. Wood-Lark, another. Linnet, one. Jack Snipe, several. Siberian Chiffchaff, a female in turnips: a sad-coloured little bird, with black legs. Garden-Warbler and Lesser Whitethroat in turnips. Greater Wheatear, one. Common Wheatears, very few. Reed-Buntings, three. Gold-crest, one. Ring-Dove, one. Merlins, several. Little Bunting, one on wall, very wild.

9th October.—South-east, strong breeze; then west, light, dull, warm; south-east later, and finally south breeze.

Redwings, many around lantern at 2 A.M., and thousands on island in large flocks. At 6 P.M. large numbers were seen to leave the island, rising high and flying south-west towards the Orkneys—a most interesting incident, for emigrants are rarely detected actually embarking on their night journeys. Thrushes, a goodly number. Blackbirds, a number of males seen on cliffs. Mistle-Thrush, one flying high and "churring." Wheatears, a few still here, including the large race. Fieldfare, one only with the Redwings; this bird never haunts the cliffs, but seeks the high heathery ground. Siskins, several. Coot, one captured at the head of a geo. Little Buntings, two seen; they joined a flock of Twites on being disturbed, and were "lost." Ring-Doves, two seen on the skerries on the south-west side, and two on land. Reed-Buntings, a number. Jack Snipe, several. Bramblings and Chaffinches, many.

10th October.—South-west, strong breeze.

Redwings, very numerous. Blackbirds, several. Single examples of Siskin, Wigeon, and Turtle-Dove.

11th October.—South-west, strong breeze.

Single House-Martin seen.

12th October.—South-west, strong breeze; much too strong for migration.

Snow-Buntings, a few making their first appearance in the crofts. Tree-Pipit, only one left; Lapp Bunting, one. Redwings, Song-Thrushes, Bramblings, and Chaffinches still here.

13th October.—South-south-west, light breeze.

Greater Wheatear last noted. Common Wheatears, several. Woodcocks, two. Wigeon, three. Song-Thrushes and Redwings decreasing. Skylarks, small party seen arriving at the North Lighthouse. Ring-Ouzels, a few. Swallow, a single bird. Blackcaps, a female, caught at south lantern at 8.15 P.M.

14th October.—South, strong breeze; raw and hazy. South-east in evening till 12 P.M.

Redwings, considerable numbers have appeared; also several Ring-Ouzels, Siskins, and Curlews. Common Sandpiper, one. House-Martin still here.

15th October.—North-west, light; at night changed to east, then south-east.

Yellow Bunting, an arrival. Little Buntings, several; two seen in turnips and others heard among the Twites. White Wagtails, two (the last) left about this date. Shore-Larks, several. Merlins seen for last time. Rook, first for 1908. Lapwings, two. Willow-Warbler, one.

16th October.—South-east till afternoon, then south; thick fog.

Between 6 and 9 P.M. Redwings were continually striking the lantern of the South Light; about a hundred seen during the day. Mergansers, three.

17th October.—With the south wind, an immigration has taken place; weather very cold.

Great Northern Diver, in the North Haven. Glaucous Gull, the first for the season, a young bird. Lapwings, seven. Lapp Buntings, three. Mergansers, several. Mistle-Thrushes, two. Goldcrests, everywhere to-day, on the cliffs and in the crofts alike, uttering their peevish little call-note. They give one the impression that they are suffering, for they sit very close and pay little attention to the presence of an observer. An example which came into the houses readily sat on the occupier's arm, and devoured about a score of the common housefly. Yellow-browed Warbler, one was observed busily engaged seeking for food in company with several Goldcrests in one of the wildest parts of the island. When disturbed, it flew to the nearest cliffs. Blackcap, a male flew into a house, and a female seen on cliff. Siskins, an addition to the number previously present. Fieldfares, several arrivals. Skylarks, many have appeared. Wheatears, a few were scattered over the island. Woodcocks, three. Redbreasts, several.

18th October.—East-south-east breeze; clear.

Siberian Chiffchaff, one searching for food on the sides of a ravine. It allowed so close an approach that its black legs could be distinctly seen. Little Buntings, no less than six came under observation, sheltering and seeking food on the sides of the same ravine. Redbreasts very numerous.

19th October.—Southerly gale; much colder.

The wind much too high for observation, and the birds were in hiding. When any of the smaller birds were found and disturbed, there was scarcely any chance of seeing them again, for the wind instantly carried them far away. Rose-coloured Pastor, an example was well seen among a number of Redwings and Song-Thrushes. It afterwards left these, and joined a flock of Starlings. It was very wary, and kept to the face of the cliff. Starlings, a large number, several hundreds of immigrants in to-day. They kept to the open fields, and did not join resident birds and visit the yards. Common Gulls, very numerous and mostly old birds. Lapwings, seven. House-Martin, one. Blackbirds, very numerous. Short-eared Owls, two. Rooks, three. Woodcocks, between forty and fifty. Jackdaw, one. Grey Geese, two. Goldcrest, only one. Blackcap, male. Little Buntings, three or four seen.

20th October.—Southerly gale continues; wet forenoon.

Birds more in evidence to-day. Goldcrests, many. Woodcock, scarcer. Scaup, one. Yellow-browed Warbler, one. Blackcap, male, still here. Common Gull, about one hundred and fifty young and old birds.

Black-headed Gulls, many with the last. Yellow Bunting, one. Short-eared Owls, three. Wigeon and Teal, plentiful. Mergansers, five. Mallard, one. Swan, one. Greenfinches, five. Glaucous Gull, an immature bird. Lapwings, fifteen.

21st October.—Southerly, but wind decreased; colder.

Redwings much scarcer. Siskins still here. Ducks plentiful. Wood-Lark, one. Lapwings increased, thirty seen. Goldcrests, many still here.

22nd October.—South-south-west, strong breeze.

Fieldfare, a dozen arrivals. Yellow Buntings, two. Richard's Pipit, one. Siskins, eight still here. Lapwings, further increase. Linnet, one. Lesser Black-backed Gull, an adult. Goldcrests scarcer.

23rd October.—South-south-west breeze.

Great Northern Diver, one. Woodcocks, numerous. Siberian Chiffchaff, one. Fieldfares, increase. Richard's Pipit, one again seen.

24th October.—South breeze; clear.

Woodcocks especially numerous to-day. Over the hills these birds could be roused from beside every little tussock and clod of earth, and many crouched by the sides of stones. Even on the crofts and cultivated land individuals were seen, and some were flushed out of cabbages. They also resorted to the cliffs, on the grassy ledges of which Woodcocks could be seen from the edge of the cliff—always sitting in one position, namely, with their tails up and their bills down. About three hundred at least were in the isle. Two Richard's Pipits were seen together. It is notable that these birds always affected the sides of the braes and open country, where it was impossible to approach them. Fieldfares, more numerous. Long-eared Owls, two. Blackbirds, very numerous; about one hundred seen. Corn-Bunting, one. Mealy Redpolls, two. Goldcrests, numerous.

25th October.—South-west, light; clear.

Shore-Lark, three arrivals. Goldcrests again very numerous.

26th October.—South-east, light breeze; clear.

Richard's Pipit - after much stalking and considerable difficulty, one was at last secured. It kept persistently to the open, and constantly took long flights, uttering its single loud note repeatedly. Little Buntings—two were put out of marshy ground, which they seem to prefer to any other. This bird is usually solitary, or at the most two or three together. Woodcocks—the flight is over; only some dozen birds were seen. Mealy Redpoll, one. Siskin, one. Coot, one. Hedge-Accentor, one. Wood-Lark, one. Common Gulls decreased.

27th October.—East-south-east, breezy.

Short-eared Owl, one. Siskins, dozen seen. Fieldfares, increase. Yellow Buntings, increase. Wood-Lark, one. A few Redwings, Song-Thrushes, Fieldfares, and Starlings captured at south lantern; also a Woodcock.

28th October.—East-south-east breeze.

Wood-Larks, three. Blackcap, a male. Woodcocks, four. Common Sandpiper, one. Ring-Ouzel, one. About 5.30 P.M. emigrating birds appeared at the lantern of the south lighthouse, and many Fieldfares, Song-Thrushes, Redwings, Blackbirds, Redbreasts, Starlings, Bramblings, Skylarks, and Chaffinches were noted until 8.30 P.M. When the weather cleared, the wind changed to the north-west, and the movement was no longer observed, if it did not cease.

29th October.—Fine, clear.

Redbreasts, numerous; Siskins still here. Common Sandpiper, one. Pomatorhine Skua, one. Richard's Pipit, one. Greenfinches, eight. Yellow Buntings, about twenty. Little Buntings, three. Lapp Buntings, two. Ring-Ouzels, two. Blackcap, female. Goldcrests, numerous. No increase of Fieldfares, Redwings, or Song-Thrushes.

30th October.—Southerly breeze; fog.

Siberian Chiffchaff, one. Hedge-Accentors observed. Redstart, one. Common Chiffchaff, remains found of one recently dead.

31st October.—South breeze; rain and fog all day.

Fieldfares, hundreds arrived about midday. Woodcocks, a few arrivals. Redwings, Song-Thrushes, Blackbirds, and Ring-Ouzels. Redbreasts and Goldcrests also appeared in numbers. Wood-Larks, two. Great Northern Divers, two. Yellow-hammers, about thirty.

1st November.—South, light; clear.

Wood-Lark, one. Red-throated Pipit, one observed at close quarters.

2nd November.—South-east breeze; foggy.

Short-eared Owl, one. Fieldfares, Blackbirds, and Goldcrests scarcer (passed on). Dunlin, one. Little Auks, a few seen off island. Common Sandpiper, one. Fulmar, one. Redstart, one. Water-Rails, two. Siberian Chiffchaff, one: this species is one of the tamest birds visiting the isle. Snipe, increase. Ring-Ouzel, one. Merlin, one. Little Bunting, two. Scaup, one. Siskins still here. Snow-Buntings, numerous. Redshanks, about a dozen seen passing over the island in a south-west direction, 10.15 A.M.

3rd November.—South-east, light.

Wood-Larks, a party of five observed. Dunlins, two. Grey Geese (? species), seven. Ring-Doves, two. Fieldfares scarcer. Wheatears, two noted.

4th November.—East-south-east, breezy.

Wood-Larks, five: these birds keep very much to the stubbles, but do not associate with the Skylarks. Fieldfares, about thirty. White Wagtail, one. Snipe, two parties of seven and four seen flying over island: these birds are generally seen in parties on arrival, not afterwards. Short-eared Owl, two. Siskin, one. Woodcocks, two.

5th November.—South-east breeze; clear.

Fieldfares, about four hundred have appeared since yesterday. Redwings and Song-Thrushes also very numerous. Slavonian Grebe, one. Goldcrests, many. Bramblings, flock which has been present for some time gone to-day. Woodcocks, about two hundred seen. Snipe, several seen on the high hills with Woodcocks. Long-tailed Ducks, a dozen. Sheld-Duck, one.

6th November.—South-east, light.

Woodcocks, a decrease noticeable. Fieldfares diminishing. Siberian Chiffchaff, one. Little Bunting, one. Jack Snipe, one. Wheatear, one. Wood-Larks, three.

7th November.—East, light.

Woodcocks, about a dozen seen, others left. Mealy Redpolls, two. Fieldfares, about fifty seen. Glaucous Gulls, several.

9th November.—Westerly, light; clear.

Little Stint, one. Tree-Pipit, one. Turnstone, one. Sheld-Duck, one. Wood-Larks, two. Little Bunting, one. Goldcrest, one.

10*th November.*—South-west breeze.

Redbreasts decreasing last few days. Fieldfares, about a hundred here. Redwings and Song-Thrushes, numerous. Blackbirds decreasing. White Wagtail, one. Yellow Buntings scarcer, about six seen. Woodcocks, four. Snow-Buntings, flocks of a hundred or more seen.

11*th November.*—Light, south; clear.

Little Buntings, two. Fieldfares, Redwings, Song-Thrushes, and Snow-Buntings scarcer; Blackbirds plentiful.

12*th November.*—Westerly, light; clear.

Carrion Crow, one. Teal, one. Fieldfares, more in. Merlin, one. Curlews, increase. Redbreasts scarcer, though still numerous.

13*th November.*—South-west, strong breeze.

Fieldfares and Snow-Buntings scarcer. Redwings and Song-Thrushes still plentiful. Goldeneye, one. Slavonian Grebe, one.

14*th November.*—South breeze; clear.

House-Martin, one. Woodcocks, five. Siskins, two. Goldcrests, several on the face of the cliffs.

15*th November.*—Nothing seen worth mentioning.

16*th November.*—South-south-west, strong.

Goldcrests, several. Song-Thrushes and Fieldfares, numerous. Redwings scarcer. Redbreasts, a few. Water-Hen, one. Carrion Crow and House-Martin still here.

17*th November.*—North-west gale; clear.

Woodcocks, four. Jack Snipe, five. Common Snipe, many since yesterday. Water-Rails, two. Goldeneye, one.

18*th November.*—West breeze; rain.

Rooks, several arrivals. Goldeneye, one. Puffin, one in North Haven. Fieldfares, Redwings, and Song-Thrushes scarcer.

19th, *20th*, *21st*, *22nd*, and *23rd November*.—Strong westerly winds, and no birds.

24th November.—Westerly strong breeze; cold.

Redbreasts, a few. Fieldfares, three or four. Lapwing, one.

27th November.—South-west breeze; clear.

Tufted Ducks, two in the North Haven. Redstart, one. Slavonian Grebe, one.

28th November.—South-south-west breeze; rain.

Short-toed Lark in company with Skylarks. Fieldfares, party of a dozen. Teal, one. Slavonian Grebe, one again seen. Ducks plentiful.

1st December.—West, light; clear.

Chaffinches, about a dozen; and Fieldfares, fourteen as arrivals. Puffin, one. Little Auks, numerous. Fulmars, two. Turnstone, one.

2nd December.—West, breezy.

Puffins, four found washed up. Jack Snipe, one. Merlin, one.

3rd December.—South-east, breezy.

Golden Plover, an arrival. Chaffinches and Snipe, a few. Redbreasts, two. Scaup, one. Blackbirds, a few. Merganser, one. Curlews, many new arrivals. Mallards, numerous. Fieldfares, twenty. No Song-Thrushes or Redwings seen.

4th December.—South breeze.

Fieldfares, about two hundred arrivals since yesterday, probably from Shetland on their way south. Blackbirds, a few more in. Snipe, several. Mallards, a dozen. Greenfinches, many.

5th December.—South breeze; clear.

Redwings, two over to-day. Rooks, two.

8th December.—South-west breeze.

Fieldfares and Common Gulls, arrivals.

11th December.—East breeze; rain.

Corn-Buntings, three arrivals. Swans, three.

12th December.—North-east, light wind; clear.

Greenfinches, a few. Redwings, four. Fieldfares, about a hundred. Skylarks, a number. Corn-Buntings, several at present on the island.

17th December.—Glaucous Gulls, many young birds, one adult.

27th December.—East breeze; clear.

Fieldfares, about fifty more in to-day. Rooks, two new arrivals; both species probably from Shetland.

28th December.—South-east breeze.

Redwings, four arrivals.

29th December.—South-east, strong; snow.

Coot and Water-Rail captured, and another of last-mentioned species seen. Dunlins, a few arrivals. Fieldfares and Blackbirds, a few still here.

30th December.—South, strong breeze; slight thaw.

Woodcock, one. Snipe, one. Dunlin, one. Mallards, several.

31st December.—Mallards, three. Teal, one. Golden Plover, one. Black-headed and Common Gulls present in small numbers. Lapwings, about seventy. Fieldfares, about a hundred.

CHAPTER XXI

THE BIRDS OF FAIR ISLE, WITH SPECIAL REFERENCE TO THE SEASONAL MOVEMENTS OF THE BIRDS OF PASSAGE

From the foregoing chapters it will have been gathered that the avifauna of Fair Isle is a remarkable one. Its wealth in migratory species, already wonderful, seems to be by no means exhausted, for every year adds either new birds to its ornis, or something of value to the data relating to the migrations of the regular visitants. In all no less than 209 species are known to have occurred on the island. Of these, 28 are natives—that is to say, breeding birds, 18 being residents and 10 summer visitants. 117 are visitors on their seasonal passages; of these, 101 are observed in spring on their way northwards, and 115 in the autumn proceeding southwards; 99 being common to both seasons. 14 are winter visitors to the isle, and 48 have occurred as casual visitors of a greater or lesser degree of rarity; of these, 23 have occurred in spring, 33 in autumn, and 10 at both these seasons. The Sea Eagle, once a native, is now, alas, a bird of the past.

Corvus corax, *Raven.*—Some fifteen years ago no less than six pairs of Ravens nested annually on the island. These had to be reduced, owing to the injury they inflicted on the lambs; and now one pair only survive. In the autumn of 1909 the family party consisted of eight birds, six young having been successfully reared, an unusually large brood.

Corvus corone, *Carrion Crow.*—There are two records of the visits of this species, namely, of single birds in the autumns of 1907 and 1908. The first of these I frequently saw during September in company with Grey Crows. The second arrived in November, and only remained a few days.

Corvus cornix, *Grey Crow.*—As a resident, this species is fairly numerous, six pairs at least nesting on the cliffs. Immigrants arrive when the autumn is well advanced, and these are said to be distinguishable from the resident birds by reason of their keeping in flocks, while the native Grey Crows hunt for food singly or in pairs. The Duchess of Bedford informs me that on 3rd October 1911, great numbers of immigrants were observed. They seemed to be all over the island, and after remaining two or three days, took their departure.

Corvus frugilegus, *Rook.*—Occurs commonly on both the spring and autumn passages, but is somewhat irregular in its appearances, being seen in numbers in some years, and almost entirely absent in others. A few occasionally occur in winter.

The spring passage generally commences during the second week of March, and is in progress throughout April, but stragglers have occurred until the end of the first week of May. In 1910, however, numbers appeared during the last week of February.

In autumn there are a few records of arrivals in September; but it is not until the latter half of October and onwards to mid-November that it may be expected, and then only in small numbers.

In spring the birds resort to the newly sown land, and are considered a great pest. George Stout noticed on two occasions in spring, when the Rooks were more than usually plentiful, that a certain kind of worm infested the land, and one Rook that was killed had its crop full of this pest, instead of the seeds which the islanders expected to find.

CORVUS MONEDULA, *Jackdaw.*—A few occur during spring and autumn, usually in company with Rooks.

In spring their appearances date from 21st March to 13th May, but chiefly during April.

The autumn visits have been made between 19th October and 10th November.

STURNUS VULGARIS, *Starling.*—An abundant resident, nesting in the cliffs and buildings. These native birds are strictly sedentary, and have chosen roosting-places amid the cliffs, assembling on the rocks towards dusk ere they retire. Among these chosen night-resorts are certain natural tunnels, of considerable length, which connect the Western Ocean with two very remarkable chasms, known as the Reevas. In the recesses of these retreats, with the surf of an ever-restless sea roaring below them, numbers of these birds pass the night: wilder and more weird roosting-places it would be difficult to conceive.

The Starling is also a common bird of passage, moving northwards in spring from the end of the third week in March until mid-April.

The first of the autumn migrants are recorded for 9th September, and they continue to arrive and pass southwards throughout October and until the fourth week of November. These birds of passage during their sojourn on the island resort to the open fields to a much greater degree than do the native birds. Many are killed at the lanterns of the lighthouses at this season.

PASTOR ROSEUS, *Rose-coloured Starling*.—Single adult males are reported, on good evidence, as having visited the island in the springs of 1907 and 1908; and a third on 19th October in the latter year.

ORIOLUS ORIOLUS, *Golden Oriole*.—Has been known to visit the island on three occasions during recent years. An adult female, on 11th May 1908; a young male, on 21st September 1908; and one, probably a male, seen 26th May 1909.

CHLORIS CHLORIS, *Greenfinch*.—The Greenfinch is an irregular winter visitor, some years being plentiful, while in others not a bird is seen. Its chief rôle in Fair Isle is, however, that of a bird of passage.

In the autumn it has arrived at dates varying from 3rd October to 26th November, the usual date of its appearance being mid-October. Some of the passage movements at this season have been of considerable magnitude.

Fewer are observed in spring, when the passages date from 18th April to 5th May. In 1910, however, twelve appeared on 5th March. The dates for its advent at both seasons are somewhat irregular.

COCCOTHRAUSTES COCCOTHRAUSTES, *Hawfinch*.—Single adult males have been known to visit the island on two occasions in spring—namely, on 14th May 1908, and on 8th May 1909. Both these birds were observed searching for food amid the dung of ponies.

PASSER DOMESTICUS, *House-Sparrow*.—An abundant resident. Inquiries instituted for the purpose of eliciting information as to when the island was first colonised completely failed in their object, and the only conclusion arrived at was that the bird had been a native, time out of mind.

PASSER MONTANUS, *Tree-Sparrow.*—This bird is a fairly common resident, and nests chiefly in the inaccessible crevices on the face of the great chasms known as the Reevas. To these safe retreats the birds retire to roost all the year round. There does not appear to be any evidence in favour of the bird being regarded as either an autumn visitor or passing migrant. It is an extremely wary species, and, though quite unmolested, is most difficult to approach.

FRINGILLA CŒLEBS, *Chaffinch.*—Common as bird of double passage, some of the autumn visitors remaining through the winter.

The earliest record for its appearance on the spring migration is for 22nd March. It passes throughout April and until the second week in May, often in considerable numbers. Stragglers occur later, and have been observed down to 8th June.

Has appeared on its return in autumn as early as 3rd September, but the passages southwards do not set in, in earnest, until the fourth week of that month, and are in full swing until mid-November. Both males and females appear in company in the flocks that arrive in the autumn.

FRINGILLA MONTIFRINGILLA, *Brambling.*—An abundant visitor during both passages, some of the autumn birds remaining for the winter.

Appears in spring on its way north from 8th April, and occurs in considerable numbers until about 18th May, but has been observed as late as 6th June. The birds seen during May have either donned, or are rapidly assuming their pretty summer plumage; and a flock of them illumined by bright sunlight presents a singularly beautiful appearance. Indeed, these are the handsomest of all the spring birds of passage.

Arrives annually in abundance in September, the 19th being the earliest and the 21st the average date for its appearance. In 1910 none were observed before 29th September. The passage continues throughout October, and smaller numbers are observed during the first half of November.

SPINUS SPINUS, *Siskin.*—A regular visitor in varying numbers in autumn, but is less frequently observed in spring.

In autumn it has appeared at dates varying from 22nd September to 14th November, and was exceptionally abundant from 28th September to 4th October in 1909.

In spring it has only come under notice in small numbers—namely, on 23rd April 1909; on 8th May 1911; on 12th May 1908; and on 13th and 21st May 1910.

ACANTHIS CANNABINA, *Linnet.*—A few birds, never more than three in company, have appeared in both spring and autumn. In spring the dates of its advent vary from 29th March to 19th May; and in autumn from 17th August to 25th October.

Linnets associate with the numerous bands of Twites, amid which they are difficult to detect, and it is possible that the birds may be more frequent and plentiful than is suspected.

ACANTHIS FLAVIROSTRIS, *Twite.*—This is a remarkably abundant species, numbers being resident; though fewer are present during the winter. In the autumn, the Twite population of the island is to be reckoned in thousands. Early in the season they are to be found everywhere, especially on the face of cliffs where seed-bearing plants flourish. Later, after the corn is cut and garnered, they gather together, form immense flocks, and frequent the stubbles. The extraordinary abundance of this species adds much to the difficulties of detecting and identifying immigrants, especially the Finches and Buntings, which on arrival usually join the ranks of the Twite armies, where they become to all intents and purposes effaced. On this account this cheery little bird becomes in the autumn a veritable curse to the bird-observer. A diminution in their numbers takes place late in the autumn. The return of the autumn emigrants takes place in early spring, and many were noted as arrivals on 28th February and 15th March 1910. The arrivals observed at this season probably include birds on passage—migrants whose appearance it has been impossible to detect in the autumn.

ACANTHIS RUFESCENS, *Lesser Redpoll.*—A male captured on 1st May 1906, is the only occurrence known; but the species has several times been recorded for Shetland.

ACANTHIS LINARIA, *Mealy Redpoll.*—Redpolls are not infrequent visitors in the autumn, and in some seasons are extremely abundant; but unless examples are obtained, it is usually quite impossible to say to which of the several races of Redpoll they belong.

It is probable that the typical form under notice is the most frequent of all, and of it specimens have been captured in March, April, and May—dates varying from 3rd of March to 2nd of June. Others, possibly of this race, have been seen as late as 16th June.

Down to the year 1910, however, I had not obtained *Acanthis linaria* in the autumn, though many "Mealy Redpolls" had appeared in the observers' reports as seen between 6th August and 27th December. In the autumn of 1910, however, *Acanthis linaria* was exceptionally abundant. The first to appear arrived on 16th October, when several were observed. More followed on the 21st, and many hundreds were present at the end of the month, after which they were seen in gradually decreasing numbers down to 28th November—the last date for their appearance in the records. The return of these hosts in the following spring of 1911 was recorded between 3rd March and 2nd June, the birds being most numerous on 8th May.

ACANTHIS HOLBOELLI, *Holböll's Redpoll.*—Among the vast numbers of typical Mealy Redpolls which visited the isle in October 1910, were several, doubtless many, which belonged to this large form. A few were sent to me as examples of the common bird, the wings of some of which measured 80 mm. I do not regard the birds forming this supposed sub-species[1] as being anything more than large individuals of the ordinary form (*Acanthis linaria*). The two completely intergrade as regards the size of their bills and wings—their sole diagnostic characters—and are found in the same areas in the northern regions of both the Old and New Worlds.

ACANTHIS ROSTRATA,[2] *Greater Redpoll.*—This large, dull-coloured, heavily-striped native of Greenland was extremely abundant during our visit in the autumn of 1905. It appeared on 17th September, became numerous by the 25th, still more so on 2nd October, and was common down to the 6th, the day of our departure

[1] *Acanthis linaria holboelli.* [2] *Acanthis linaria rostrata.*

from the island. All the specimens obtained, and the scores of them carefully examined with the aid of binoculars, were in juvenile dress—that is to say, none of them showed any signs of rose-colour on the breast. Some of them fraternised with the hordes of Twites, and sought food in the stubbles; while others moved about in large parties, and frequented the enclosures near the houses, being attracted by the seeds of numerous weeds which abounded there. This bird, like the Twite, is much given to bathing about midday, and was always to be found thus engaged at that hour on certain little streams. The Greater Redpoll appeared again in small numbers on 9th September 1908, when it remained until 7th November, and again in October 1911.

ACANTHIS HORNEMANNI, *Greenland Redpoll*.—Of this rare and beautiful native of Greenland, Iceland, Jan Mayen, and Spitzbergen, not less than five occurred during my visit in the autumn of 1905. The first to come under notice were a party of three, consisting of an adult male and two younger birds, which appeared on 18th September. These birds frequented an enclosure in front of one of the crofter's houses, where they fed on the seeds of weeds for several days, and were exceedingly tame. On the 29th a second adult male was observed seeking food among some low herbage; and on 10th October another young bird was found. In life these birds, especially the adults, appeared to be almost entirely white, and this fact, and their habit of puffing out their fluffy feathers, rendered them exceedingly pretty and conspicuous objects.

ACANTHIS EXILIPES, *Hoary* or *Coues Redpoll*.—During the great autumn invasion of Mealy Redpolls, which formed the outstanding feature in the ornithological retrospect of 1910, three immature specimens of this small race of the last species were obtained. Doubtless many others were present, for those captured were not sought for, and were unknown to their captor. These birds were a female obtained on 26th October, a male on 3rd November, and one of doubtful sex on 5th November. This race[1] has hitherto proved to be an extremely rare visitor to the British Isles, having only occurred in the counties of Yorkshire and Hertfordshire. Its native haunts are in Northern Russia, Siberia, and Arctic America.

PYRRHULA PYRRHULA, *Northern Bullfinch.*—A party, consisting of about a dozen birds, of both sexes, of this Northern European and Siberian Bullfinch, appeared on 5th November 1906, and remained some days. Another and more remarkable visitation occurred in the autumn of 1910, about the same time as the hordes of Mealy Redpolls appeared. On this occasion the birds were first seen on 24th October, and some remained practically all the winter on the isle, since small numbers are recorded as being present down to 16th December.

During these visits the birds were chiefly seen in the cultivated enclosures near the houses, but also here and there in all parts of the island, and were very tame.

CARPODACUS ERYTHRINUS, *Scarlet Grosbeak.*—This was until quite recently supposed to be one of the rarest of the visitors appearing in the British Isles. The first Scottish specimen, a bird of the year, was shot at Fair Isle by Kinnear, from a patch of potatoes, on 3rd October 1906. On 4th September 1908 an adult female was obtained; and on the 29th an immature male was

[1] *Acanthis hornemanni exilipes.*

found among a flock of Twites on the stubbles. A scarlet bird reported and described to me during my visit in the autumn of 1908 was, no doubt, an adult male of this species.

Loxia curvirostra, *Common Crossbill.*—During the remarkable irruption of Crossbills from the Continent in the summer of 1909, Fair Isle received many of the visitors. These birds were first detected on 23rd June, and their numbers increased gradually until 10th July, as if the birds had arrived in a series of waves, as many as 300 being seen on a single day. The immigrants remained on the island throughout July, but their numbers fell off towards the end of the month. They were observed in some abundance, in small scattered parties, during the whole of August. Later they became gradually scarcer, only two or three being seen in September; and the last of the hundreds once present was seen on 2nd October, feeding on the head of a thistle. During their sojourn they frequented all parts of the island: the faces of the great cliffs, the cultivated land, the grassy slopes, and the high heathery ground. On the latter they fed on the unripe fruit of the crowberry; elsewhere on the seeds of grasses and other plants, and on the heads of thistles. At first they seemed to thrive, but later many perished, their dead bodies being found in the plots of potatoes, etc.

A small number made their appearance in the summer of 1910, the first of these visitors being an adult male seen on 20th June. None appear to have been seen afterwards until 8th August, when nine came under notice, and the birds were seen daily in small numbers down to 19th September.

Another immigration took place, in the summer of

1911, when the birds were seen from 22nd July to 11th September, twelve being the most for any day.

LOXIA BIFASCIATA, *Two-barred Crossbill.*—This native of the North Russian and Siberian pine forests has occurred on two occasions. An adult male on 13th June 1908, and another on 10th July 1909. The latter was feeding on the ground amid a number of examples of the common species, then so abundant. It is more than probable that there were other examples of this species on the island at the time, which escaped notice, females and young birds being particularly liable to be overlooked among the throng of the commoner bird.

EMBERIZA MELANOCEPHALA, *Black-headed Bunting.*—A female example of this summer visitor to South-Eastern Europe occurred on 21st September 1907. It had been feeding on Phalangids, and on the seeds of grasses. A second specimen, this time a young male, was obtained, 25th August 1910. These are respectively the sixth and seventh known visits of this species to the British Islands.

EMBERIZA MILIARIA, *Corn-Bunting.*—This species nested on the island in 1905, when we found young birds; but does not appear to have done so since, until 1911. It is, however, a regular visitor on passage in spring and autumn. It has also occurred now and then in winter.

In spring it has been recorded from 11th March, throughout April, and as late as 22nd May.

In autumn its passage visits date from 11th September to 29th November, but never more than three have been observed on any day.

EMBERIZA CITRINELLA, *Yellow Bunting.*—Is regularly

observed on both the spring and autumn passages, and has been seen occasionally in winter.

The first birds of spring appear between 21st March and 8th April, and the passage north lasts until 23rd May.

The return autumn flight sets in during the second week of September, and is in progress until the end of the first week of November. In 1910 it was unusually numerous from 17th October to the end of the month.

Emberiza hortulana, *Ortolan Bunting*.—The Fair Isle investigations have proved this bird to be a regular and fairly common visitor to the British Isles during both its spring and autumn passages.

The earliest date for its arrival in spring is 30th April, after which it appears at intervals until as late as 6th June, and a single bird was observed on 23rd June 1910. It is most regular and numerous, however, during the middle weeks of May.

In autumn the dates for its appearance range from 29th August to 18th September, and the passage south lasts throughout the month. In 1907 it was observed down to 9th October.

During its visits in the spring, the Ortolan frequents newly sown land, where the young corn is beginning to appear. It is then much less shy than in the autumn. At the latter season it is very partial to standing corn, to which it retreats when disturbed elsewhere. Later it is to be found feeding on the stubbles.

Emberiza rustica, *Rustic Bunting*.—There are three known instances of the occurrences of this species at Fair Isle—namely, an adult female on 18th June 1908; a bird of the year on 22nd September 1908; and an adult male on 29th September 1909. Two of these visitors were found amid rough grass, while the last-mentioned example was seen feeding on a stook of oats.

EMBERIZA PUSILLA, *Little Bunting*.—Until the Fair Isle investigations were instituted, this species was regarded as one of the rarest casual visitors to our islands. Through them it has been proved to be a bird of double passage, and not uncommon in the autumn.

The earliest dates on which it has been observed in the fall range from 18th September to 10th October, and the passage lasts until 11th November. It is never numerous, but as many as six have been seen on a single day; and these probably represented only a small proportion of those present.

It has been observed on three occasions in spring—namely, on 14th April 1907; on 12th May 1908; and two on 18th May 1909.

In the autumn it chiefly resorts to the stubbles, where it finds congenial company in the hordes of Twites, amid which it is most difficult to detect. The call note is not unlike that of the Yellow Bunting, but is shorter and fainter. When taking flight the tail is spread, and the white outer tail-feathers are then conspicuously displayed.

EMBERIZA SCHŒNICLUS, *Reed-Bunting*.—Is a bird of double passage, occurring regularly in small parties at both seasons.

The earliest date for its appearance in spring is 2nd March, the latest 6th June, and the average 4th April. The period covering its main movements is from 9th April to 19th May. The numbers which arrived on 12th May 1910 and succeeding days were simply phenomenal. There were hundreds of both sexes engaged in seeking food on newly-sown land, and others were similarly engaged on the rocks at the foot of the cliffs, where they were observed to take short flights in the pursuit of flies. At night these birds retired to the grasslands and fallows to roost, some of the parties containing quite as many as one hundred individuals. The birds participating in this most remarkable passage movement remained on the island until the 21st, and a few lingered until the 27th.

The autumn passages date from 20th September to 20th November, the main period ranging from 28th September to 30th October.

This is the only Bunting that seeks cover in the crofts, where it is frequently found among the patches of potatoes in the autumn.

PLECTROPHENAX NIVALIS, *Snow-Bunting.*—Common as a winter visitor and bird of passage. It makes its first appearance in autumn between 8th and 18th September, when a few are to be seen on the high ground; and arrives in numbers and visits the crofts early in October and November. In 1911, however, a single bird was seen on 29th August. During snow these birds form large flocks, and come quite close to the doors of the houses in their search for food.

The winter visitors, and others from the south which have arrived during March, leave in the first week of April. The birds on passage appear about the same date, and the flights northward are witnessed until the second week of May, though stragglers have been observed as late as 14th June.

CALCARIUS LAPPONICUS, *Lapland Bunting.*—Occurs annually as an autumn visitor; and has during several seasons come under notice on its return passage in spring.

The dates for this bird's appearance in autumn range from 25th August to 9th September. It is not uncommon during the latter month and in the first week of October in most years, while the latest date for its presence is 29th October. At this season it is to be found in the wilder portions of the island, and chiefly among rough grass, on the seeds of which it feeds; but after the corn is cut it often seeks the stubbles.

In spring it appears to be much scarcer, but has come under notice on five occasions, between 25th March and 2nd May. An adult male in summer plumage was observed on the last-named date. On its spring visits it has been chiefly detected among the hordes of Twites, and on account of the company it then keeps is very liable to be overlooked.

ALAUDA ARVENSIS, *Skylark.*—The Skylark is found at all seasons of the year, but observation has demonstrated that the native birds are summer visitors and quit the island in the autumn, their place being taken later by visitors from the North, some of which remain the entire winter.

The home birds leave early in September, and the first foreigners appear soon after, from about the 19th and onwards to the early days of November (the 5th latest), often in vast numbers. Only a small number remain for the winter.

The return of the native birds commences as early as the end of February. In March the arrivals are pronounced, and are continued, on the part of birds on their passage north, throughout April.

ALAUDA ARBOREA, *Woodlark.*—Though this species

has only twice been detected elsewhere in Scotland at the time of writing, yet it arrives annually in the autumn at Fair Isle, but is not abundant, though small parties are not uncommon. The dates between which its first appearance has been detected range from 7th October to 5th November, and it has some years been observed throughout winter. It has been observed in spring as late as 2nd April.

During its sojourn it frequents all parts of the island, but is most frequently seen on the stubbles and other land under cultivation. It is noted that Woodlarks very rarely associate with the Skylarks, from which they can easily be distinguished by their flight and notes.

CALANDRELLA BRACHYDACTYLA, *Short-toed Lark.*—This uncommon straggler from Southern Europe has occurred on not less than three occasions. The first visitor, a male, was detected by its lighter colouring and smaller size, amongst some Skylarks on 11th November 1907; the second, a female, was found on 18th June 1908; and, lastly, a male was obtained on 28th November 1910. These Fair Isle birds vary much in colour, but I have been unable to assign them to separate races.

OTOCORYS ALPESTRIS, *Shore-Lark.*—This bird occurs annually in autumn, and sometimes remains well into winter. It has, as yet, only once been observed in spring.

The dates upon which it has first appeared in autumn range from 15th to 23rd October, after which arrivals have taken place as late as 17th November. It has been observed in winter down to 16th December; and the spring bird was seen on 6th March.

It usually frequents the high heathery ground and

hill-sides, but has also been observed on the stubbles. It is generally very shy, and difficult to approach.

MOTACILLA LUGUBRIS, *Pied Wagtail.*—Occurs regularly on passage in small numbers in spring, when it is chiefly found on newly-ploughed land.

In autumn it is much less in evidence, and some seasons has entirely escaped notice.

The spring records for its appearance vary from 3rd to 25th March; and the latter half of this month and down to mid-April covers the period during which it has been mainly observed. It has, however, several times been seen in May, and even as late as the 28th of that month.

Its few visits in autumn have occurred between 22nd July and 1st October; the majority of them have been in August.

MOTACILLA ALBA, *White Wagtail.*—This is one of the most abundant birds of passage visiting the island, and it occurs regularly in both spring and autumn. A pair reared their young in the summer of 1909, and again in 1910 and 1911.

The dates for its appearance in spring range from 5th to 20th April, but the first three weeks of May cover the main period during which the passages at this season are performed. A few appear as late as the first week of June.

In autumn the bird has been observed as early as 9th August, but the return movements do not set in, as a rule, until the middle of that month, and last in force until mid-September. It is, however, observed annually in small and gradually decreasing numbers down to mid-October; and one was sent to me which had been procured on 10th November 1908.

In spring the migrants are chiefly observed following the plough; but in autumn their favourite resorts are the few beaches, where they capture the flies which abound in the belt of seaweed which fringes high-water mark.

MOTACILLA BOARULA, *Grey Wagtail.*—There are as yet only five known instances of the occurrence of this species—all in spring, and, with one exception, of single birds. In one instance two appeared. This species can only, it is thought, be regarded as a straggler beyond its accustomed range, since it has no summer haunts in Northern Europe, being there, as in Fair Isle, a rare visitor only. The dates of these visits are as follows: 11th, 18th, and 20th April, and 17th and 21st May.

MOTACILLA RAYI, *Yellow Wagtail.*—This is another bird whose visits to Fair Isle have carried it beyond the range of its usual habitats, which do not extend beyond the southern half of the Scottish mainland. It has occurred on four occasions, all in spring: an adult male on 8th May 1906; a male on 11th May 1908; two on 23rd April 1909; and one on 27th April 1909. The specimen obtained on 11th May 1908 was an abnormally pale one.

MOTACILLA FLAVA, *Blue-headed Wagtail.*—Four of these birds appeared on 18th May 1908, and an adult male was captured and sent to me.[1] It had been previously recorded for both the Orkney and Shetland groups, but the occurrences of this species and of the hitherto much overlooked *M. thunbergi* (= *M. borealis*)

[1] On the 14th of May 1910, and again on the 18th and 19th, adult males appeared, and for several days came under the notice of the Duchess of Bedford and myself. These birds haunted the grasslands, and showed a strong predilection for the company of cattle.

have been confounded in the past, so that it is now impossible to decide to which form the older records should be assigned.

MOTACILLA THUNBERGI,[1] *Grey-headed Wagtail.*—A regular visitor in small parties in spring, and in scantier numbers in autumn.

The earliest record for its appearance in spring is for 25th April; but the dates between which it most frequently occurs are from the second week of May to the end of the month. The latest date is for 3rd June, when several appeared in 1907.

In autumn it has been noted between 26th September and 9th October, and once on 4th November.

It is an extremely wary species and very difficult to approach, except when engrossed in capturing flies disturbed by cattle.

ANTHUS TRIVIALIS, *Tree-Pipit.*—Common on both passages, though apparently overlooked elsewhere in the Shetland area.

The earliest date for its appearance in spring, on its way northward, is 5th May. It is numerous from the 12th of that month to first days of June; and the latest date for its appearance is 10th of the last-named month.

The return autumn movements are timed from 26th August, and are frequent from the middle to the end of September; a few occur in October, sometimes as late as the 22nd; while in 1908 one was sent to me which had been obtained on 9th November.

The chief haunts of the spring migrants are the grassy faces of the cliffs, on which they obtain food in company with various Warblers. Though sometimes frequenting the same ground, it seldom keeps company

[1] *Motacilla flava thunbergi.*

with the Meadow-Pipit, and is usually a shyer bird. In autumn it is often flushed from patches of potatoes.

ANTHUS PRATENSIS, *Meadow-Pipit.*—Strange to say, this species is not a native bird, though the island abounds in suitable nesting haunts. It is, however, very abundant on both the spring and autumn passages. Odd birds have been known to appear as early in the year as 27th January and 14th February, and a few regularly arrive in March. It is not until the third week of April, and onwards to the middle of May, that it is observed passing in numbers, and the 19th of the last-named month is the latest date on which it has been noted.

It is more numerous still in autumn, and 29th August is the earliest date on which it has appeared. It is abundant throughout September, and occurs commonly down to mid-October. There are no November records, but a single bird appeared on 11th December 1906.

ANTHUS CERVINUS, *Red-throated Pipit.*—I had paid careful attention to the Pipits, and examined many thousands through my field-glasses, in the hope of detecting this species, which is one likely to occur annually on passage at Fair Isle. It was not, however, until 2nd October 1908 that success was achieved, and then my attention was drawn to the bird by its peculiar note as it passed high overhead. I was fortunate enough to secure it, and it proved to be a female of the long-looked-for bird. Its note was not at all like that of any Pipit or other bird with which I was acquainted. Another was heard on 1st November of the same year.

ANTHUS OBSCURUS, *Rock-Pipit.*—An abundant

resident. After the nesting season it is very common in all parts of the island, especially on the stubbles after harvest. I have been unable to ascertain to what extent this species is an emigrant or an immigrant, but suspect that it is both.

This bird is one of the greatest of feathered bullies. No sooner does a Warbler, Redstart, or other small bird appear in the vicinity of a Rock-Pipit's preserves than it is attacked, pursued, and driven away. The bully, in this case, is not a coward, for I have seen it assault Starlings and Redwings as readily as a diminutive Goldcrest.

ANTHUS LITTORALIS, *Scandinavian Rock-Pipit.*—I saw a fine male of this race[1] on 16th May 1911. My attention was drawn to it by the vinous tinge on its breast, which was quite conspicuous as it stood facing the sun. It is a native of Scandinavia, whence some, at least, move south on the approach of winter.

ANTHUS RICHARDI, *Richard's Pipit.*—Several of this summer visitor to Central Europe appeared on the island in the autumn of 1908. It was first noticed, as a shy stranger, about 26th September; but its identity was not solved until the 30th, when a male was shot. In October several more were detected between the 22nd and 29th, and specimens obtained. In 1909 one was seen on 26th September, and in 1911 two came under the notice of the Duchess of Bedford and Wilson, the observer, on 6th October.

These birds frequented the crofts and open grass-lands, and were extremely wary, and hence difficult to approach.

CERTHIA FAMILIARIS, *Creeper.*—On 27th December

[1] *Anthus obscurus littoralis.*

1906, a male was discovered in an exhausted state, and allowed itself to be captured while endeavouring to find shelter under a stack of hay. It had most probably been blown across the North Sea from Scandinavia by the fierce gales which prevailed just previous to its appearance. The bird belonged to the typical form inhabiting Northern Europe and Asia, not to the British race *Certhia familiaris britannica*, and is, I believe, the only example of the Continental race known to have visited our shores.

PARUS MAJOR, *Great Titmouse.*—A female example of the Continental Great Titmouse (which is to be distinguished from its British cousin mainly by its more slender bill) was obtained on 17th November 1910. Saxby, in his delightful book, "The Birds of Shetland," mentions two previous visits for Shetland, one of which occurred in Unst in April, and the other in Yell in the early autumn.

LANIUS EXCUBITOR, *Great Grey Shrike.*—Has appeared occasionally, but always singly, on both passages.

In spring its visits have been confined to April, and the dates upon which it has been observed range from the 6th to the 26th of that month.

The autumn appearances range from 18th October to 9th November.

On the 9th of November 1907, one was shot while in hot pursuit of a Rock-Pipit. This bird belonged to the variety which possesses only a single wing-bar.

LANIUS COLLURIO, *Red-backed Shrike.*—Occurs regularly on both passages, but adults have only come under notice in spring.

The spring visits date from 11th May to 2nd June.

During these movements the bird is not uncommon, and both sexes were particularly abundant from the 12th to the 23rd of May 1910, arriving concurrently.

In autumn young birds only have hitherto been detected, and the bird has always been somewhat scarce at this season. The passages have been confined to September, and are recorded from the 9th to the 20th of the month.

Although it appears in some numbers in spring, it is usually seen singly, and frequents stone walls, fences, and even the face of the cliffs.

In May 1910, I saw these Shrikes pursue and capture the fine, locally distributed Bee, *Bombus smithianus*, which is abundant on the island; and I have taken an entire example of this insect from the crop of a young bird in the autumn. George Stout once disturbed one which was feeding on the remains of a Rock-Pipit; and has seen it chase other small birds.

REGULUS REGULUS, *Goldcrest*.—The Continental race of Goldcrest is an annual visitor on passage during both the spring and autumn; but is much more in evidence and abundant during the fall movements.

In spring, on its journey northwards to its summer haunts, it has appeared as early as 25th March; but the average date for its advent at this season is 6th April, and the movement lasts until the end of the first week of May.

In autumn the earliest-known visitor was observed on 8th September, but the average date for its arrival is 19th September. In most seasons it arrives in numbers during the fourth week of September, and in varying abundance at intervals throughout October. In some seasons many have been recorded as late as the

5th of November, after which only a few have been observed, and the 15th is the latest date chronicled for its presence on the island. Its habit of frequenting the face of the great cliffs in search of food has previously been alluded to. It is also to be found amid potatoes and turnips in the autumn.

SYLVIA NISORIA, *Barred Warbler.*—This species has so far escaped detection on its passage movements in the spring. The records of its autumn visits have not been numerous, and relate to the occurrence of twelve single examples—adults and young. It is probable, however, that it often escapes notice.

The earliest date for its appearance is 3rd August, and the latest 16th September. There are, however, only four known instances of its occurrence in August (3rd, 8th, 16th, and 26th). It has usually been found amidst cover about unoccupied houses and other secluded places, and has sought the face of the cliffs when disturbed.

SYLVIA SYLVIA, *Common Whitethroat.*—Though abundant on its spring flight northwards, this species is much less so on its return journey in the autumn.

The earliest date for its first appearance in spring is 4th May, the latest 10th May, and it occurs throughout the month and during the first week of June. On two occasions Whitethroats have been seen on the exceptionally late dates of 18th and 25th June—probably birds that would never reach their nesting haunts.

With one exception all the autumn records are for September, and on dates ranging from the 5th to the 30th. The exceptional instance was a single bird on 3rd October. Its chief haunts in spring are the sides of ditches, old lichen-covered walls, and the face of the cliffs. In the autumn, when cover is to be had, it has been found amid standing corn, and not unfrequently among cabbages.

SYLVIA CURRUCA, *Lesser Whitethroat.*—Much more abundant and frequent in appearance than the last-named species. Indeed, it is, along with *Phylloscopus trochilus*, the commonest of the Warblers visiting the island.

In spring its first appearances date from 23rd April to 9th June; the average date for its arrival on passage being 7th May. It has been seen as late as 20th June —probably a laggard, non-breeding bird.

In autumn the earliest record of its return is 20th August, the average 25th August, and the latest 17th October. It is frequent throughout September, and has many times occurred during the first week of October.

In the autumn this little bird is very partial to the cover afforded by the plots of potatoes, turnips, and cabbages, especially the latter. In these it lies very close, and when disturbed it darts forward with a jerky flight, during which it usually spreads its tail, and reveals its identity among Warblers by the display of the whitish outer tail-feathers. In spring the sides of ditches, the large boulders above high-water mark, and the face of the cliffs are its chief resorts.

SYLVIA ATRICAPILLA, *Blackcap.*—Has appeared in spring on its northward journey at dates ranging from 28th April to 15th June, but is not much in evidence until after mid-May, and then onwards until the first week of June. The average date for its first appearance is 11th May.

The autumnal visits commence late in August, the

24th being the earliest date for its advent, and are in progress throughout September and during the first half of October—the latest date for its presence being 29th October. In 1908 both sexes were numerous on 24th and 25th October. Some males in October have their heads clad half in black and half in reddish-brown feathers.

Like most of its congeners, it seeks cover on arrival in the autumn, and must be sought for amid the shelter afforded by plots in the cultivated area.

SYLVIA BORIN, *Garden-Warbler.*—The earliest record for this bird's arrival on its spring passage is 24th April, and from 9th May to 10th June it is seen in small numbers only—chiefly during the middle weeks of May.

It is quite abundant, however, in the autumn on its way south. Its visits at this season, usually commencing during the first week of September, continue throughout the month, and until the 11th October on the part of stragglers. In the autumn of 1910, an early season, the first immigrants were seen on 17th and 27th August. As in spring, single birds are observed; but these are now frequently widely distributed, and hence quite numerous.

It autumn it is decidedly skulking in its habits, and is not easily put on the wing when detected among cover, such as potatoes, which are its favourite haunt. In spring it seeks the scanty shelter afforded by the withered grasses of the ditches. It also frequents the cliffs at both seasons.

SYLVIA SUBALPINA, *Subalpine Warbler.*—The second British example, an adult male, occurred on 6th May 1908—a great day for migratory visitors to Fair Isle (see Diary of Movements). That the second known visit to our islands of this pretty little Warbler should, like the first, have been to one of the most remote spots within the British area, seems on first thoughts somewhat strange. It must be remembered, however, that in such corners as St Kilda and Fair Isle, should there be anyone there to notice them, waifs like these are more likely to be detected than elsewhere. If this bird, which is a summer visitor to Southern Europe and Northern Africa, has found its way to such distant parts of the British area, it must surely have occurred elsewhere with us.

PHYLLOSCOPUS SIBILATRIX, *Wood-Warbler.*—The few Fair Isle records for the visits of this species, and the single visit to Sule Skerry, are the only instances known to me which suggest the occurrence of this species as a bird of passage in the British Islands. Until the year 1910, two single birds only had been observed in spring, namely, on 6th June 1907, and 5th May 1909. On 12th May 1910, quite a number arrived, and chiefly resorted to the face of the cliffs, and were excessively confiding, for several that came to the top sought insects at my feet. Here these birds, or other arrivals, continued to be observed until the 18th.

For the autumn there are only two records for its appearance, namely, on 2nd August 1907, and 21st August 1911. It is probable that during recent years the breeding range of this species has extended westwards of the Christiania Fjord, and that this may explain the bird's visits on migration to the British Heligoland.

PHYLLOSCOPUS TROCHILUS, *Willow-Warbler.*—As one would naturally expect, the Willow-Warblers obtained on migration at Fair Isle belong to both forms which seek the North for their summer haunts. Thus we have both *Phylloscopus trochilus eversmanni* and typical *T. trochilus* as visitors; the former being the North-Eastern European and Siberian race, and the latter the Scandinavian.

This bird is one of the most abundant of the smaller migrants visiting the island on its spring and autumn passages.

In spring it arrives from the south at dates between 8th April and 6th June, but is most frequent throughout May, and occurs in rushes down to the end of the third week of that month. Vast numbers appeared between 12th and 15th May 1911, and all, with one or two exceptions, belonged to the race *eversmanni*.

Its autumn movements also cover a considerable period, since they have been chronicled between 7th August and 25th October, the main passages covering the entire month of September. In the autumn of 1910, an early season, it was numerous during the last week of August, and less abundant than usual in September. A few occur annually in October, chiefly during the first half of the month.

During its visits it is to be found almost everywhere. In spring it is one of the few sylvan migrants that are sufficiently cheerful to indulge in a few notes of song; and in the autumn I have frequently heard it utter its pretty, plaintive call-note.

PHYLLOSCOPUS BOREALIS, *Northern Willow-Warbler.* —When searching for migratory birds on the 28th September 1908, I disturbed, in a patch of potatoes where it was hiding, a Willow-Warbler with very dark upper plumage. This I was fortunate enough to secure, though with difficulty, for it was extremely wild. It proved to be a male, and its stomach contained a Phalangid, two Lepidopterous larvæ, and the remains of various Dipterous insects. At first this bird was thought to be new to the British avifauna, but I afterwards found that the Warbler recorded for Sule Skerry as *P. viridanus* belonged to this species. The Northern Willow-Warbler does not appear to have occurred since in the British Islands, and has only once, I believe, been obtained elsewhere in Western Europe, namely, at Heligoland, on 6th October 1854. The bird, however, is a summer visitor to Finmark, Northern Russia, and Siberia, but does not seem to have any winter quarters in the western portion of the Old World, its cold-weather retreats being confined to Southern Asia.

PHYLLOSCOPUS ABIETINA, *Northern Chiffchaff.*

Chiffchaffs appear on their passages regularly in both spring and autumn, and belong to this large and long-winged race[1] which, among other Northern European countries, summers in Scandinavia, to and from which, no doubt, the visitors to Fair Isle are travelling. A number of specimens have been obtained whose wing measurements are as follows:—Males, from 62 to 64 mm.; females, from 56 to 64 mm.

The earliest of the spring migrants was killed at the lantern of the South Lighthouse on 7th May, and the passage northwards has been observed as late as 1st June, the main period being from 13th to 20th May.

Its appearances in autumn are rather late, the earliest date being 26th September, from which date down to 24th October it has occurred at intervals.

[1] *Phylloscopus collybita abietina.*

It frequents much the same haunts as the Willow-Warbler, and has the same habits, but is always silent.

PHYLLOSCOPUS TRISTIS, *Siberian Chiffchaff.*—This is the latest of all the Warblers to make its appearance at Fair Isle, where it has appeared regularly during the autumns of the past four years. It has, as yet, only once been detected on its passage northwards in spring, namely, on the 7th of May 1909.

The earliest dates on which it has been known to appear during each of these four seasons are on 8th October in 1908, 15th October 1910, 21st October 1907, and 30th October 1909. The latest date on which it has been observed in the island is 6th November. In all about a score have been obtained; but it is not very uncommon in most seasons. I only once had the pleasure of seeing the bird in life, namely, on 8th October 1908, my visits on other occasions having come to an end ere it arrived. George Stout, who has seen quite a number of them, considers this to be the tamest species of *Phylloscopus* visiting the isle. He found the birds chiefly in the cabbage-plots, the only cover in the late autumn, where they were sometimes seen in company with the Northern Chiffchaff. On some occasions he saw several together; and on fine days they perched on the cabbage leaves, every now and then darting into the air to capture insects, and returning to their perch. They were also observed searching for flies on the face of the cliffs, on peat banks, and on old lichen-clad walls. Elsewhere in Britain, it has only been known to occur at Sule Skerry, in Orkney in the winter of 1907-8, and at the Isle of May on 16th October 1910.

PHYLLOSCOPUS SUPERCILIOSUS, *Yellow-browed Warbler.* —This is a favourite with me; I made its acquaintance at Fair Isle during my visit in 1905, and have seen it each autumn I have returned to the island, as many as seven examples on a single day. It is a regular visitor on autumn passage in small numbers; but since it is very partial to retreats on the face of the great range of lofty cliffs on the west side of the isle, it is doubtless more abundant than our knowledge permits us to assert.

The earliest record for its appearance is 19th September; but it usually arrives during the fourth week of that month, and has occurred each year during the first half of October. The latest date for its presence on the island was on 29th October in 1907.

In addition to the retreats and feeding-ground afforded by the face of the cliffs, this bird also visits the crofted portion of the island, where it is chiefly to be found in the cover afforded by the potato plots.

HYPOLAIS HYPOLAIS, *Icterine Warbler.*—The first specimen known to have occurred in Scotland was observed on some bare, open ground on 1st June 1908. It appeared amid a rush of birds, and proved to be a female. Between the 3rd and 5th of June 1911 no less than three appeared. Although this species is a summer visitor to Scandinavia, it has only been occasionally found singly on our shores during the periods of its passage movements. As the bird is not uncommon in the north-western portion of the Continent of Europe, its migration routes to and from Norway presumably do not traverse our British shores.

ACROCEPHALUS SHŒNOBÆNUS, *Sedge-Warbler.*—Although this is a regular and fairly abundant visitor on its spring passage, yet it has only twice been detected in autumn, and then only single birds were observed.

The vernal movements commence with an observation for 7th May, from which date it has occurred down to 10th June; but it is most in evidence during the middle weeks of May.

Regarding the autumn, the single birds were detected on 19th and 20th September in 1906, and both frequented the standing corn—a fact which may account for our want of success in observing it, for standing corn cannot be searched. I know from experience on the Yorkshire coast that this species is partial to such cover when on migration. In spring its chief haunts are the sides of ditches.

ACROCEPHALUS STREPERUS, *Reed-Warbler.* — Has occurred, presumably as a straggler, on four occasions in the autumn: all these visitors were single birds and came under my personal observation. Strange to relate, these are the only known instances of the Reed-Warbler's appearance in Scotland; and are also, I believe, the most northerly records of its occurrence in Europe. It seems to be quite unknown in Norway, but visits South Sweden, and is common in Denmark.

The following are the particulars relating to the Reed-Warbler's visits in Fair Isle, all of which are for September, and at dates ranging from the 24th to the 30th of that month. On 24th September 1906, a male shot from potatoes.[1] In 1908, one found in standing corn on the 24th September, and another in turnips on the 26th. In 1909, on 30th September, a female was obtained on an oat-stook.

[1] I am now of opinion that this bird is a Marsh-Warbler (*A. palustris*) in first plumage.

ACROCEPHALUS DUMETORUM, *Blyth's Reed-Warbler.*—On the 29th of September 1910, a small Warbler was found by the Duchess of Bedford in a patch of turnips. It only afforded a very brief view ere it flew off. It was present in the same cover on the following day, and then aroused suspicion that it might perhaps be something uncommon. After a great hunt it was secured, and sent to me.

In colour it more resembles the Marsh-Warbler (*A. palustris*) than the Reed-Warbler (*A. streperus*), but is a little duller in colour and usually smaller in size, though the smallest Marsh-Warbler and the largest *dumetorum* overlap. The wing formula of Blyth's bird differs from both the British species just mentioned in having the second primary shorter than the fifth.

Blyth's Reed-Warbler is a new bird for Western Europe, not having previously, I believe, been found nearer than Russia. It is a summer visitor to regions from St Petersburg and Archangel eastwards to the Yenisei (Western Siberia), and the Himalayas (Kashmir to Nepal); and winters in the plains of India, in Ceylon, and from Sind to Assam.

LOCUSTELLA LUSCINIOIDES, *Savi's Warbler.*—The visit of a pair of these birds to Fair Isle must be regarded as one of the most remarkable events in the annals of British ornithology for many years. These birds appeared on the 14th May 1908, and frequented the grassy sides of a small burn. They were extremely shy, and for several hours baffled all attempts to secure one of them for the purposes of identification. At last, however, one, a female, was obtained and its unlooked-for indentity revealed. This bird was once a summer visitor to the fens of East Anglia, but ceased to be a native species more than fifty years ago. Since then it has not been known to occur, even as a casual visitor, in the British Islands. That it should have reappeared among us at a locality so far removed from its former English haunts, is most remarkable. Even Heligoland, with its unrivalled record of feathered waifs, cannot boast of having Savi's Warbler amongst its long list of distinguished visitors. Fair Isle is the Ultima Thule of its known wanderings in any land; and, needless to remark, it had never been known to visit Scotland before.

LOCUSTELLA NÆVIA, *Grasshopper-Warbler.*—This species has been known to occur on two occasions in spring—namely, on the 14th of May 1908, when two appeared; and on the 29th of May 1907, when a female was killed during a rush of birds.

So far as I am aware, this species is of rare occurrence in Norway; and its appearance in Fair Isle may possibly be attributed to its having either overshot its summer range when seeking its breeding haunts in our islands, or been carried out of its course by unfavourable weather conditions.

LOCUSTELLA LANCEOLATA, *Lanceolated Grasshopper-Warbler.*—On the 9th September 1908, I secured a bird of the year of this species, as it rose from some rough grass. It was new to Scotland, and had only twice previously been recorded for Western Europe, namely, at North Cotes in Lincolnshire on 18th November 1909; and at Heligoland on 13th October 1909. A second Scottish example was caught at the Pentland Skerries on 26th October 1910.

This species is a summer visitor to the whole of Siberia and the northern Isles of Japan, but is of very rare occurrence in European Russia. Its winter quarters are in Burmah, India, South China, and Borneo. This eastern bird resembles our Grasshopper-Warbler, but, as a rule, it is a little smaller, and has the dark spots on the centre of the feathers of the upper plumage more clearly defined. The Orkney specimen, which is a bird of the year, has the under parts greyish yellow, with dark brown streaks on the centre of the feathers of the throat and breast.

TURDUS VISCIVORUS, *Mistle-Thrush.*—A few appear on both the spring and autumn passage movements to and from Northern Europe. In spring it has been observed between 6th March and 12th May; and its autumn visits date from 1st October to 20th November, but at this season it is chiefly observed during the latter half of October. Usually only one or two birds are observed on each occasion, but most probably a number escape notice, as the bird chiefly frequents high ground in the more remote portions of the island.

TURDUS MUSICUS, *Song-Thrush.*—Birds belonging to the Continental race (the typical *Turdus musicus*) occur on both passages in considerable numbers, and a few of those arriving in the autumn pass the winter on the isle.

The spring movements northwards commence on the part of a few birds late in March (22nd earliest); but their arrival in numbers takes place from the end of the first week of April to that of May. Later, stragglers or small numbers occur, and have been recorded down to 8th June.

In autumn the first immigrants usually appear in mid-September, but there are no great arrivals until the second week of October, and these are witnessed at intervals until 21st November. At this season the turnips are the main resort of the immigrant Thrushes, but many resort to the face of the cliffs.

Sometimes during snow a few arrive along with Fieldfares and Redwings. These, no doubt, are birds which have passed the previous portion of the winter in Shetland, and are moving south under the pressure imposed by severe weather conditions.

In the summer of 1905 a pair reared their young in the face of a ravine, where I saw young birds just able to fly early in September. In 1911 the bird again nested in the island.

TURDUS ILIACUS, *Redwing.*—Occurs regularly on both passages in considerable, sometimes vast numbers. A few appear during the prevalence of severe winter weather, coming from the north, but do not remain.

It has been observed on its way north in spring, as early as 23rd March; but as a rule the passage does not set in until the end of the first week of April, and lasts until the end of that month. A few are sometimes seen early in May, the 15th being the latest date on which they have been observed. In 1906 a single bird appeared amid a rush of migrants on 1st June—an exceptional occurrence.

The average date for first arrivals in the autumn is 25th September, the earliest 19th September; and it occurs in numbers, often in great rushes, down to the third week of November. The cliffs seem to be the chief resorts of this species during its sojourn on the island, but it is also very common in the shelter afforded by turnips.

TURDUS PILARIS, *Fieldfare.*—An abundant species during its spring and autumn migrations. Some also appear in winter, during spells of severe frost or snow, along with Thrushes, Redwings, and Blackbirds, but such visitors only remain a short time on the isle.

The earliest date for its appearance in spring is 22nd March, but it is not until the second week of April that it may be expected. From this date down to mid-May it occurs in numbers, with stragglers to the very end of the month.

In autumn I have twice observed single birds in September—namely, on the 23rd and 30th, but the earliest date for its appearance in small numbers is 2nd October. In 1909 no arrivals were detected until 18th October. It is abundant between the latter half of October and the end of the third week of November.

During its visits to the island it generally frequents the high open ground; and although it occasionally seeks the crofts, it never enters cover like the Thrush, Redwing, and Blackbird; nor does it haunt the cliffs.

TURDUS MERULA, *Blackbird.*—Like the other members of its genus, the Blackbird is a common species of double passage. A few, after arrival in the autumn, spend the winter in the isle.

The spring movements towards the north commence soon after mid-March (18th earliest). Many have appeared late in the month (24th), and the passages are in progress until near the end of April. Stragglers have occurred during May, and even in early June (5th latest).

The average date of its advent in autumn is 26th September (14th earliest). It is not, however, until after 12th to 18th October that it becomes more or less abundant, and it continues to be so at intervals down to 29th November in some years.

Its retreats during its sojourn on the island are the same as those of the Song-Thrush.

TURDUS TORQUATUS, *Ring-Ouzel.*—The Ring-Ouzel

is a regular visitor on both the seasonal passages, and is not at all uncommon.

The earliest date on which it has appeared in spring is 18th March, but the next earliest record is not chronicled until 8th April. It has not been seen in numbers before 23rd April, nor after 17th May, but stragglers occur down to the 21st. Exceptionally it has been observed as late as 15th June.

The average date for its return is 28th September, (earliest the 23rd, latest 5th October). In some seasons it is numerous during the first half of October, but is not usually so until the 18th and after; and the passages have been observed down to 2nd November, but a single bird was seen on the 27th in 1910.

During its sojourn it frequents the faces of the cliffs and chasms, also the stone walls of the higher and more remote parts of the isle.

Erithacus rubecula, *Redbreast*.—The Continental race of the Redbreast is a common and regular visitor throughout its spring and autumn movements to and from its North-Western European summer haunts. A few of the autumn migrants pass the winter in the isle.

The spring passage commences during the second week of April, and laggard migrants have occurred as late as 23rd May. The main period for the vernal movements is between the second week of April and the end of the first week of May. A single bird was observed on 9th June 1910, and several were seen on 10th March 1908.

The dates of first appearance in the autumn range from 20th to 26th September. It is numerous from 18th October to 8th November, and smaller numbers arrive down to the 24th.

When on passage this is an extremely wild bird, and is not to be seen in the neighbourhood of the houses. It is mostly then to be found away from the crofted portion of the island, and is especially partial to the water-mills, in which the natives grind their corn, which are situated in lonely parts of the isle. Those that remain all winter, however, become more confiding as the season advances, and then approach the houses in their search for food.

LUSCINIA LUSCINIA, *Thrush Nightingale.*—This was one of the rarities which rewarded us during our visit in the spring of 1911. As a mere waif, it arrived along with a crowd of migrants on the 15th of May—a day on which no less than twenty-four species of birds of passage came under our notice. It was observed creeping about the rocks on the shore at the foot of a cliff behind the south lighthouse, and proved to be a male.

This species is larger than the Common Nightingale, from which it also differs in having its upper surface darker and more olive-brown in tint, the tail dark brown and only slightly reddish, the chest and sides of the breast darker and having a mottled appearance, the first primary very short and narrow, and the third the longest feather in the wing.

Its summer range extends from Denmark and south and mid-Sweden to S.W. Siberia, and it winters in East Africa.

This is the second record, but the first satisfactory instance, of the occurrence of this species in the British Isles.

CYANECULA SUECICA, *Arctic* or *Red-spotted Bluethroat.*—This species occurs regularly, but in varying numbers, during its autumn passages, being in some seasons fairly abundant and in others little seen—perhaps because it escapes observation on the face of the cliffs. During the last four springs—*i.e.*, those of 1908, 1909, 1910, and 1911—it has occurred in some numbers.

The spring visits date from 24th April to 2nd June, and as many as a dozen have been seen in a single day. In 1910 the Duchess of Bedford and myself found the bird in fair numbers from 14th to 25th May, and probably not a few escaped our notice on the face of the cliffs.

The autumn movements are remarkable for the constancy of the dates between which the bird's first appearance has been recorded, which range from 20th to 25th September. The period covering the passage is a singularly short one, for there are only three records of its occurrence in October, all for single birds, and the latest for the 9th of the month.

Some notes on the habits of this species will be found under 23rd September on page 84.

CYANECULA CYANECULA, *White-spotted Bluethroat.*—Two males in full summer plumage have occurred in the spring. It is a matter of uncertainty whether any female examples have visited the island, as they are indistinguishable from those of the preceding species.

The first of these interesting and beautiful visitors appeared on 22nd March 1909, an early date; and the second occurred on 14th May 1910, and frequented the side of a small burn. This latter example was under the observation of the Duchess of Bedford and myself for some time. Its actions were identical with those of a Redbreast; in disposition it was shy, and when disturbed flitted away, but always returned to the same haunts. Several birds of the Red-spotted species also arrived at the same time.

These occurrences are the only known visits of this species to Scotland, and Fair Isle marks the northern limit of the known wanderings of this Central European summer bird.

RUTICILLA PHŒNICURUS, *Redstart.*—The Redstart is a regular and common visitor in both spring and autumn.

The earliest dates for its advent in spring are 22nd March 1909, and the 4th and 9th April, when single males appeared. There are other April records for the 18th, 21st, and 23rd; but it is not until 12th May that it has appeared in numbers, and the passage usually lasts until the end of the third week of the month. In some seasons a few have been observed as late as the end of the first week of June, and a single bird was seen on 13th June in 1910.

According to the records, the males arrive first in spring, being from six to seventeen days in advance of the females. Later both sexes appear simultaneously on passage.

The return movements usually commence in September, the 13th being the average date for the bird's appearance. In the autumn of 1910, an early season, the first migrant was seen on 29th August. The main flights occur towards the end of September, but small numbers appear during October, and there is one record for 2nd November.

Its haunts during its visits are the stone walls in the vicinity of the houses. Its main retreats, however, are the high cliffs of the western side, where insects are abundant. Here it may be seen on fine days ever and anon darting out from the precipitous faces in pursuit of flies, its identity always being clearly indicated by the display of its "fire-tail."

RUTICILLA TITYS, *Black Redstart.*—There have been six known visits of single birds of this species to the isle during as many years. Four of these have been made in the springtime, and two in the autumn: they are the most northerly records for the British Islands.

The spring records are for 9th April 1907, when a male in the plumage of *R. carii* appeared; a male in full summer plumage on 20th May 1910, a similar bird on 12th May 1911, and one, sex unrecorded, in a rush of migrants on 31st May 1908.

The autumn birds were both females, and the dates of their visits were 14th October 1909, and 8th November 1907.

I saw two of these birds, and their actions on the ground exactly resembled those of the Redbreast and Bluethroats.

SAXICOLA ŒNANTHE, *Wheatear.*—A considerable number of Common Wheatears are summer visitors to the isle, whilst many more occur as birds of passage in spring and autumn.

The earliest dates for its appearance in spring are 13th, 21st, and 28th March. It occurs regularly in the first week of April, and is abundant on passage from the 7th of the month down to the early days of May, the 12th being the latest date for its passage en route for the north.

The males appear first, and are recorded as being from five to fourteen days in advance of the females. Even during the later movements the males often greatly outnumber the females.

Many of the native birds and their young leave the isle before the end of August. The migrants from Northern Europe pass southwards throughout September, and are much fewer in numbers during the first half of October; while stragglers forming the rearguard have been seen as late as 14th November—when a single bird was observed.

During their visits the migrants on passage are widely scattered over the higher ground and the unenclosed parts of the island.

An albino example was obtained on 29th August 1911.

SAXICOLA LEUCORRHOA, *Greater Wheatear.*—This fine race[1] of the Common Wheatear occurs in considerable numbers in both spring and autumn on its passages to and from the north-west—Iceland and Greenland.

The earliest date on which it has been observed in spring is 3rd May, when numbers appeared, and the passage lasts until 1st June. In this species the males appear to arrive a day or two earlier than the females.

In autumn it returns late in August, and has been observed until mid-October (16th latest). In the late autumn it is found in some numbers after nearly all the representatives of the common species have passed southwards.

SAXICOLA HISPANICA, *Black-throated Wheatear.*—On the 25th September 1907, I was fortunate enough to detect a fine male of this handsome species among scattered examples of the Common and Greater Wheatears, which had arrived in considerable numbers. The contents of its crop consisted of insect matter, among

[1] *Saxicola œnanthe leucorrhoa.*

which were the larvæ of noctuid moths and the remains of several ichneumon flies.

This bird is the only one known to have visited Scotland,[1] and there are only a very few instances of its occurrence in England. It is a native of South-Western Europe, spends the summer season there, and in the autumn seeks cold-weather retreats in West Africa.

PRATINCOLA RUBETRA, *Whinchat.*—A visitor on migration at both seasons; but is one of those species which are more numerous in spring than in autumn.

The spring movements are witnessed from 6th May to 6th June. The average date for its appearance is 9th May, and the bird is very constant in the observation of the time of its coming.

The autumn passages have been chronicled from 24th August, and are in progress throughout September. There are several records of its visits in October, the 9th being the latest. A few birds only are seen simultaneously at this season, and the tops of the sheaves of corn are their favourite perches.

PRATINCOLA RUBICOLA, *Stonechat.*—This species appears annually in small numbers in spring; but there are at present only three records for visits in the autumn, all for single birds.

In spring its appearances date from 7th March to 27th May. The autumn records are for 8th and 26th September and 22nd October. In spring it usually appears singly or in pairs, but on 22nd March 1909, five or six were seen in a rush of migrants.

The Stonechat has, I believe, only once been recorded elsewhere for Shetland, and it is not known to have occurred in Norway. The object of its regular spring

[1] I have since obtained it at St Kilda.

visits to Fair Isle are inexplicable; nor is it possible to determine satisfactorily whether the Fair Isle Stonechats belong to Dr Hartert's British or the Continental form, as the few specimens captured were obtained in spring.

ACCENTOR MODULARIS, *Hedge-Accentor.*—The Hedge-Accentor which visits Fair Isle regularly on passage in the spring and autumn is not of the British race, but a native of Continental Europe. This bird, the typical *A. modularis*, is to be distinguished from its British cousin, *A. modularis occidentalis*, by its paler mantle, whiter abdomen, paler and pronouncedly striped flanks, more slender bill, and by its having the second primary much longer than the seventh. It is much commoner on the spring passage than on that of the autumn.

The earliest date for its arrival in spring is 29th March, but the usual date for its advent is during the first week in April. The chief period within which its passages are performed at this season is between the end of the first week in April and the middle of May, after which it is less numerous, and the latest date for its appearance is 29th May.

On the 9th of April 1907, remarkable numbers were present in all parts of the isle; and on 27th April 1908, unusual numbers again appeared.

In autumn the earliest date on which it has been found is 25th September, and the latest 16th November. It does not appear in any numbers until the last week of October, and stragglers only have been seen after the first week of November.

ACCENTOR COLLARIS, *Alpine Accentor.*—On the 6th of October 1908, I saw a bird of this species, at close quarters, resting on the face of a great cliff flanking a geo on the west side of the isle. I might have shot it with ease, but had no desire to drop it into the Atlantic surf, which laved the base of the precipice several hundred feet below. Eventually it flew off and was lost amid fastnesses so vast and unapproachable as to render pursuit quite impossible. This bird is the only example which has, as yet, been detected in Scotland.

TROGLODYTES TROGLODYTES, *Wren.*—This is a fairly abundant resident, its chief haunts being the faces of the cliffs, the ravines, and the stone walls. The only definite evidence of the visits of immigrants are the observations made by the Duchess of Bedford, who noticed a decided increase in the numbers of the bird on 25th September 1910, and 3rd October 1911—days when other migratory birds were also observed as arrivals.

The native birds seem to me to be larger than the typical form, and more nearly resemble *T. t. zetlandicus*, and probably belong to that race.

MUSCICAPA GRISOLA, *Spotted Flycatcher.*—Is annually observed on its spring migrations northwards, and is in some years very abundant at that season. On its return southwards, however, it has not been detected annually, and never in any numbers. It is another of those species which occur chiefly in the spring.

The spring appearances date from 12th May to 19th June, and rushes have been recorded between 12th May and 1st June.

In autumn the records of its visits have been between 15th September and 4th October—a late date, but one vouched for by the capture of the specimen.

On May 12th, 13th, 15th, and 22nd, 1910, great numbers appeared and frequented the face of the cliffs, the beaches where decaying seaweed abounds, and the low rocks and reefs at the south-west corner of the island. Here, when the sun exerts its genial influence, flies are numerous. During rain and dull weather many were seen on the newly-sown sodden land, apparently endeavouring to find something wherewith to sustain life. Such are the shifts to which some birds are put during certain stages in their great migratory flights: no wonder many perish.

MUSCICAPA ATRICAPILLA, *Pied Flycatcher.*—This species occurs in both the spring and autumn, on its way to and from its northern summer haunts, but is only observed in comparatively small numbers. Unlike the last species, it is most abundant in the autumn.

In spring its visits range from 6th to 17th May; while those of the autumn have been noted between 3rd and 27th September. At the latter season it is a frequent visitor to the lanterns of the lighthouses, and is often killed.

During its sojourn, it frequents the cliffs, fences, and small enclosures about the houses. All the old males have assumed their winter dress ere they appear in the autumn, and I have never seen a bird either at Fair Isle or elsewhere, during my many autumn pilgrimages in search of migrants, which showed the slightest trace of the pied plumage which renders it such a beautiful and conspicuous object in spring.

MUSCICAPA PARVA, *Red-breasted Flycatcher.*—This unusual visitor to such a northern station occurred several times during the autumn of 1906, and once in that of 1907. I was present on these occasions, and saw the birds and obtained specimens. There is only one record for spring.

The following are the particulars of the visits:—On

20th September 1906, three or four were seen. They were very wild, but an adult and a young bird were captured. On the following day two were seen; and on 4th October, one, a bird of the year, was obtained. In 1907, on 27th September, I saw one very distinctly; but it, too, was very wild, and sought the recesses of the face of the cliffs on being approached. On 31st May 1908, George Stout had an excellent view of an adult male.

I was much puzzled by the appearance of the first example that came under notice. It was a bird of the year, and hence in plain and unattractive plumage. I found it on the face of a comparatively low cliff, and my attention was especially drawn to it from the fact that it raised its tail to such an extreme degree as to be almost parallel with its back. In flight, the white on the basal portion of all but the two central tail-feathers is displayed and is very conspicuous.

HIRUNDO RUSTICA, *Swallow.*—The Swallow is another of those species which, contrary to the general rule, is more abundant in spring than in autumn.

In spring it often appears in considerable rushes, and its passages cover a long period. The earliest date for its advent at this season is 17th April; but the bird does not occur in any numbers until the end of the first week of May, after which many appear at intervals until the end of the first week in June, and in smaller numbers to the end of the month. No doubt a number of the Swallows visiting the isle at this season are non-breeding birds. A few have appeared in July, but there are as yet no August records.

The autumn visitors consist of small parties or single birds, both old and young, which arrive at intervals in September and during the first half of October.

The Swallows which arrived in numbers on 19th May 1910 seemed to be tired out, for they frequently settled on the grass close to the observers, and also on the rocks at the foot of the cliffs.

HIRUNDO RUFULA, *Red-rumped Swallow.*—On the 2nd of June 1906, the isle was the scene of the arrival of a great number of migrants rushing northwards. These included Bramblings, Tree-Pipits, Spotted Flycatchers, Red-backed Shrikes, Swallows, Cuckoos, Dunlins, and other species. Along with these were three examples of the interesting bird under notice, which is a species not previously known to have visited the British Isles, its summer home being in South-Eastern Europe. George Stout noticed these birds for several days, his attention being attracted to them by the red band across the lower portion of their backs. He succeeded in shooting one, which was unfortunately lost, but was found some ten days afterwards and forwarded to me for identification.

A single bird of this species was obtained at Heligoland on 30th May 1885, and one has since occurred on the south coast of Kent, namely, on the 16th May 1909.

CHELIDON URBICA, *Martin.*—Like the Swallow, this species is most frequent and numerous during the spring passages, and is, in most seasons, little more than a straggling visitor in the autumn.

The earliest date for its spring appearance is 7th May, and the average date 15th May. It often occurs in considerable numbers late in May and during the first week of June, and a few are observed down to near the close of the latter month. During their short sojourn they are chiefly to be seen flying continuously along the face of the cliffs, especially those on the west side of the island.

In autumn single birds, or at most two, are seen simultaneously. The dates between which they have been observed range from 4th September to 14th November.

COTILE RIPARIA, *Sand-Martin.*—Every year a few appear in spring, at dates ranging from 14th May to 26th June; but on 19th May 1910, many were seen. There are only three autumn records—namely, for 23rd July, 26th September, and 9th October.

There are but a few odd records of its visits to the Shetlands; and it would seem that this species reaches Norway, where it is common in summer, by lines of flight which do not extend to the Northern Archipelago.

DENDROCOPUS MAJOR, *Great Spotted Woodpecker.*—The only year since these investigations were instituted in which this species has appeared was 1909. In the autumn of that year a small number appeared on 2nd September, and were observed until 3rd October. One or two specimens were obtained, all of which belonged to the typical Continental race. I often saw these birds during their sojourn, and they always appeared to be very ill-pleased with their lot on Fair Isle, since they frequently gave voice to their feelings by a peevish cry, usually as they ascended the flagstaff or the equally disappointing posts of the wire fences. They soon forsook such unprofitable haunts and sought the faces of the cliffs, where they doubtless fared much better, though they must otherwise have found such unusual retreats far from congenial. Some were seen searching for food on the ground, and when thus engaged they progressed by a series of short jumps. Most of the visitors that came under close observation were birds in immature plumage.

IYNX TORQUILLA, *Wryneck*.—Although we found the remains of a number of dead birds in various parts of the island in early September 1905, yet down to the spring of 1910 a few single birds only had come under notice on their spring and autumn passages, and this not annually.

The spring visits have been made between 8th May and 4th June. In 1910, from 12th May to 24th, a considerable number were observed by the Duchess of Bedford and myself during a remarkable and almost continuous rush of migrants. Fair Isle is no place for birds with such special food requirements as the Wryneck; and those which came under notice during this memorable visitation were observed seeking for food in the newly-sown fields and on grasslands, and it is not surprising that several were picked up dead.

The autumn observations relate to appearances of single birds between 28th August and 3rd September—a very short period; but the records are few, perhaps, because it is a bird which may be readily overlooked on stone walls, its chief resorts, on which it is not easily detected—its plumage blending admirably with the colour of its resting-place.

CUCULUS CANORUS, *Cuckoo*.—Here, again, we have a species, which, contrary to the general rule, is more numerous and frequent during its spring visits than in those of the autumn.

In the spring it occurs on passage annually, usually in small numbers, at dates ranging from 7th May to 10th

June. In 1910 it appeared at intervals in exceptional abundance from 13th May to 23rd, with single birds down to 3rd June.

In autumn there are three records only for the years 1905 to 1909 inclusive, and these relate to the occurrence of single young birds observed on 20th August and 1st and 2nd September. During August 1910, however, there are five records of visits from the 16th to 26th.

Cypselus apus, *Swift.*—Swifts appear annually on both passages, but, like the species of Hirundinidæ and some others, are more numerous in springtime than in autumn. In 1910, however, the honours seem to have been pretty evenly divided between the seasons.

The earliest record for its appearance in spring is on 8th May—a very early date for Shetland—the next in order being for the 23rd. After the last-named date the bird as a rule is most in evidence, usually in small numbers, until the end of the month. Occasionally many appear in rushes with other species. There are numerous appearances for June and July on the part of a few or single birds.

In autumn one or two arrive at intervals during August and the first half of September, but I saw a single bird on 3rd and 4th October in 1909.

Caprimulgus europæus, *Nightjar.*—There are, as yet, only two records for the visits of this species. On 24th May 1910, one was observed on heathery ground by the Duchess of Bedford; and on 21st August 1908 the remains of one which had quite recently perished were found.

Upupa epops, *Hoopoe.*—There are three known instances of the visits of single birds of this beautiful species. An adult male appeared on 9th September

1907, and another on 21st October 1910. In the spring of 1910 one arrived on 13th May, and came under the notice of the Duchess of Bedford. During their sojourns these visitors were observed either on stone walls, or seeking for food on the grasslands.

NYCTEA NYCTEA, *Snowy Owl.*—There is only one record of the occurrence of this species, namely, of a bird seen on 26th October 1907.

ASIO OTUS, *Long-eared Owl.*—One or two, probably more, have appeared on several occasions in the autumn, usually in rushes with other species of migrants. These visitors have been chronicled as arriving at dates ranging from 17th October to 13th November.

There are three records for the return spring passage, all of single birds. The first of these was seen on 6th May 1908; the second, by the Duchess of Bedford, on 15th May 1910; and the third on 23rd May 1911. The second bird was resting on the face of the cliffs—a haunt where it is most difficult to detect such migrants.

A specimen obtained in the autumn of 1910 is much greyer in plumage than any other I have examined.

ASIO ACCIPITRINUS, *Short-eared Owl.*—This species appears regularly on both passages, but is most frequent and abundant in the autumn.

The spring movements northwards are recorded as occurring between 25th March and 30th May. The May visits after the middle of the month have been few.

In autumn one has arrived as early as 5th August, probably a Shetland summer bird quitting its native haunts; but 28th September is probably the first date for the appearance of Continental visitors. It has not, however, been seen in numbers until after the middle of October, and has been noted as late as 8th November.

It is usually found among rough grass, but sometimes it seeks shelter among the plots of potatoes and turnips in the crofted area.

FALCO PEREGRINUS, *Peregrine Falcon.*—There is only one pair of resident Peregrines, and these are probably the lineal descendants of the falcons which have been famed for several centuries. No doubt visitors from more northern areas appear at the island, but it has hitherto been impossible to discriminate between these foreigners and the native birds.

In the autumn the Fair Isle Peregrines prey chiefly upon young Herring Gulls, but many Woodcocks fall to them when on passage. I have seen one clutch a Ring-Ouzel on the wing and carry it off without alighting.

Writing in 1700, the Rev. John Brand, in his "Brief Descriptions of Orkney and Shetland," tells us that it is said that the Hawks "which are to be had at Fair Isle, are the best in Britain, which use to flee to Zetland or Orkney for their prey, these being the nearest lands, and sometimes they'll find Moor Fowls in their nests, which they believed to bring from Orkney, seeing there are none in Zetland, and the nearest isle they could have them in, were Stronza or Westra, which is between 40 and 50 miles of sea, over which at one flight they must carry these fowls to their nests."

FALCO ÆSALON, *Merlin.*—The Merlin has been observed in all the months of the year except June and July. It does not nest on the island, and its visits during the winter have been few. It is mainly seen at the seasons of passage.

In spring it has been observed as early as 1st April, and as late as 20th May; and in the autumn as early as

14th August. It is most frequent in September and the first half of October—the time when the small migratory birds are abundant on their southern passage, and it makes sad havoc in their ranks. On some days several are to be seen, but as a rule one or two only come under notice.

FALCO TINNUNCULUS, *Kestrel.*—The Kestrel is regular in its appearances as a bird of double passage. It has only once been recorded in winter, and has not been known to nest.

The spring visits date from 26th February, and in most seasons it is seldom seen after April. In May 1910, however, a number were observed between the 14th and the 28th. Single birds have been noted as late as 15th June.

In some years a few have appeared late in July and during August, which were probably visitors from the Shetlands; but it is much more in evidence later, when I have seen several on the wing simultaneously. The autumn movements commence with September, and the passage extends over two months. As many as six were seen on 1st November 1906, and it has occurred singly down to the 12th of that month. The only recorded instance of its appearance in winter was on 4th December 1907.

PHALACROCORAX CARBO, *Cormorant.*—Is quite common as a winter visitor. It arrives in considerable numbers at the end of August, and remains until the early spring. It has not been known to nest.

PHALACROCORAX GRACULUS, *Shag.*—A very abundant resident, finding extremely congenial haunts amid the numerous caves and in the crannies afforded by the extensive range of cliffs.

SULA BASSANA, *Gannet.*—A considerable number are to be observed fishing in the tideway, just off the island, throughout September. I have seen them thus engaged when it was nearly dark. A few are still present at the end of the first week in October, but only one or two are occasionally observed in November and December. Small numbers are seen at intervals in the spring—five were noticed on 22nd May 1910.

ANSER ANSER, *Grey Lag-Goose.*—"Grey Geese" are recorded annually on their autumn passages, and sometimes in the spring. It is only now and then that one has been obtained, owing to the open nature of the ground they frequent. Thus our information regarding the dates of occurrence of the various species which have appeared is meagre in the extreme.

On 3rd November 1908, a party of seven Grey Lags arrived, and fortunately their identity was satisfactorily established. On 9th October 1911, two came under the notice of the Duchess of Bedford.

ANSER ALBIFRONS, *White-fronted Goose.*—The wing of a bird of this species, which had been shot in the autumn of 1905, was sent to me for determination.

ANSER BRACHYRHYNCHUS, *Pink-footed Goose.*—During my visit in 1909, a party of eight of these birds appeared on 7th October, from which one was shot. I afterwards identified others which arrived on the 12th and 18th; on the latter date fourteen were observed.

BRANTA LEUCOPSIS, *Bernacle Goose.*—Single birds were seen on several occasions between 8th January and 1st March in 1909. The head of one shot in January was forwarded to me as proof of identification.

BRANTA BERNICLA, *Brent Goose.*—On 25th May 1910, four were observed, and on 27th two. The only known occurrences for the autumn are for the 16th October 1909, when an injured bird was captured in the North Haven, and 17th and 21st November 1911.

CYGNUS BEWICKI, *Bewick's Swan.*—Swans appear on passage regularly in spring (April) and in autumn (October and November), but as yet only two have been obtained, both of this species. One was shot on 18th November 1910, and the other on 30th November 1911.

TADORNA TADORNA, *Sheld-Duck.*—An occasional visitor. Single birds have been seen late in May, once in August, several times during October, and once in November. It is common in Orkney, and breeds in southern Shetland.

ANAS BOSCAS, *Mallard.*—This bird is chiefly seen during the periods of its passages in spring and autumn, but visits in winter are not unusual.

The spring migrants appear at the end of the first week of April, and are in evidence as late as mid-May. Once or twice single birds have been seen in June and occasionally in July and August.

September and down to end of November is the main period for the autumn passage of Mallards towards more southern winter quarters.

DAFILA ACUTA, *Pintail.*—This species was not detected until 1911, when, on 22nd April, a male and female appeared, followed by another bird on 19th May. In the same year, single birds arrived on 4th and 5th October, and two on the 11th of that month.

MARECA PENELOPE, *Wigeon.*—This species is fairly numerous and regular in its appearances on migration during spring and autumn, and a few are seen as visitors during the winter.

At the first season it arrives on its way north late in March, and its passages have been observed in some seasons to extend until 12th June.

It returns during the last week in August, and its visits are in progress until the third week in November. It is most abundant in October.

NETTION CRECCA, *Teal.*—Teal are fairly common during their spring and autumn migrations, but their appearance in winter is unusual, and is only once recorded.

The spring voyagers appear from the first week of March; their movements northwards continue throughout April and the first half of May; and single birds have occurred as late as 16th June.

The autumn passages are observed during the months of September, October, and November, when the bird sometimes occurs in considerable numbers. The earliest record for its appearance at this season is for 31st August, but at this early date the birds have probably come from Shetland.

FULIGULA FULIGULA, *Tufted Duck.*—This species has only rarely been observed, for Fair Isle offers it no hospitality, suitable haunts being practically non-existent. The records for its appearances are only four in number, and relate to the visits of single birds, except in one instance when two were seen. These visits were made on 28th April, 12th May, 29th October, and 27th November. They are interesting as affording evidence of passage.

FULIGULA MARILA, *Scaup.*—A few have been seen at sea in the immediate vicinity of the isle, and more rarely on the fresh-water lochs, during the spring and autumn, when proceeding to and from their boreal nesting-haunts.

In spring, single birds or pairs have been observed late in March, during April and May, and down to the 31st of the last-named month.

In autumn, in like manner, odd birds have appeared annually from 22nd September to 9th November, and one has been recorded for 3rd December.

FULIGULA FERINA, *Pochard.*—A female obtained on 2nd October 1911, is the first Pochard known to have visited the island. Another occurred on the 21st.

CLANGULA GLAUCION, *Goldeneye.*—As in the case of the Tufted Duck, only a very few, chiefly single birds, have been known to visit the Isle, all in the autumn. The dates of these visits range from 28th October to 28th November. From 16th to 18th November 1910, three were present.

HARELDA GLACIALIS, *Long-tailed Duck.*—This hardy sea-duck is common all the winter just off the island, and many doubtless occur there as birds of passage.

The earliest date recorded for its appearance in the autumn is 29th September, when a single adult male was seen. It has been observed in some seasons early in October, and it continues to arrive until the fourth week of November.

It is not unfrequent in Fair Isle waters during May, and in 1910 one was noted on 28th June.

This bird, George Stout informs me, comes close to the small boats fishing off the island. He has seen them in great numbers, but has never known them to associate with any other ducks, except Eiders.

SOMATERIA MOLLISSIMA, *Eider-Duck.* — The Eider is quite common throughout the winter; and also, as a nesting species, in summer. Eiders are, however, most numerous during the late autumn, when many appear from more northern climes on their way southwards.

A number of ducks, with broods only about one-third grown, are always to be seen in September. During the same month many of the adult males are in a variety of piebald plumages—*i.e.*, in various mixtures of post-nuptial and winter dress.

On 21st March 1911, about three hundred were seen in a flock.

SOMATERIA SPECTABILIS, *King Eider.*—I saw an adult male on the wing crossing the south bay on 13th May 1910.

ŒDEMIA NIGRA, *Common Scoter.*—A few single birds have come under notice on the following occasions:—A male on 22nd May 1910; and males on 5th and 9th September 1907.

ŒDEMIA FUSCA, *Velvet-Scoter.*—Has thrice been known to occur. An adult female was obtained on 1st December 1906; an old male on 17th September 1907; and two examples on 20th October 1910.

MERGUS SERRATOR, *Merganser.*—This species is not unfrequent, and seeks the heads of geos and the bays, where it is to be seen during both the seasons of passage, and where a few also remain all winter.

In spring, arrivals have been recorded at mid-March, but the bird is most frequent during May. It has once been observed as late as the 14th of June.

The earliest record for its appearance in autumn is for 3rd September, and the migrants continue to pass throughout the month and during October, usually in small parties of from three to five, but sometimes singly.

ARDEA CINEREA, *Heron.*—The Heron is observed regularly in both spring and autumn as a bird of passage. It also appears, but somewhat rarely, in winter.

In spring its visits are most regular during the latter half of March and throughout April, but its numbers on these occasions are small—single birds, or at most three, being usually observed. It is seen irregularly in May, and in June as late as the 28th.

In autumn it is much more numerous, and I have seen a party of as many as thirty on the wing simultaneously — namely, on 12th September 1906. A few visit the Isle in July, more in August, most in September, a few in October, and sometimes in November, and one on 13th December 1910.

On the 11th of September 1906, three appeared in the rays of the south lantern at 10.30 P.M., and remained for over an hour, croaking unmusically all the while. Several times they came close up to the light, and once were within an ace of striking, but recovered themselves just in time to avoid injury.

There are few suitable feeding-grounds for these birds, and hence they do not remain long; single birds, however, have been seen in the same haunts for several days.

COLUMBA PALUMBUS, *Ring-Dove.*—Occurs regularly on both passages, but is not numerous, eight being the largest number of migrants seen together. When disturbed whilst feeding on the land, these birds usually betake themselves to the cliffs, where the Peregrine alone can persecute them.

In spring the northern passage visits have been recorded from 23rd March to 16th June, but are most frequent late in April and during May.

In autumn it has appeared from 25th September to

10th November, the middle weeks of October being the main period for its observance.

Columba œnas, *Stock-Dove.*—This species was added to the avifauna of the island by the Duchess of Bedford. Single birds were seen on 25th, 29th, and 30th September 1911 (perhaps the same individual on each occasion), and another on 24th October.

Columba livia, *Rock-Dove.*—The Rock-Dove was formerly a resident species, finding ideal haunts in the numerous caverns and niches in the extensive range of cliffs. Until some fifteen or sixteen years ago it was abundant; now it is quite unknown. It was subject to some persecution. For instance, numbers were captured in certain caves at night. To accomplish this a sail was placed over the entrance, and then a lantern was lighted in the cave, and the frightened birds dashed against the obstruction in their endeavours to escape, and were secured. The caves, however, where this could be practised were few, and George Stout is of opinion that the introduction of guns during the construction of the lighthouses was the cause of the bird's extermination.

Turtur turtur, *Turtle-Dove.*—Has occurred annually for several years during both passages, but in small numbers.

In spring the dates for its appearance range from 9th to 27th May. Not more than two have been observed on the same date. One arrived on the 20th of June 1908, and remained until the 27th, but this was an exceptional occurrence, as was also the visit of one on 7th July 1910.

Autumn records relate to single birds which have

appeared from 16th September to 10th October; but its visits after the close of September have been few.

COTURNIX COTURNIX, *Quail.*—I was shown a clutch of eleven eggs which had been taken in July 1905. The bird breeds not unfrequently on the mainland of Shetland.

CREX CREX, *Corn-Crake.*—A summer visitor, several pairs nesting, and also a bird of double passage.

Has arrived in spring from 30th April, and has been observed in autumn as late as 18th October. The records relating to both seasons are many, but it is impossible to say which relate to native birds, and which to travellers. On 15th May 1910, I saw one apparently quite at home amid grass which was flooded with an inch or two of water.

RALLUS AQUATICUS, *Water-Rail.*—Some pass the winter in the Isle, others occur on their passages to and from the north. The earliest date for its appearance in autumn is 21st September; and in spring it has been seen down to 25th March. From the skulking nature of its habits, however, it largely escapes observation. It is not unfrequent on the reefs fringing the tide, where it seeks shelter under rocks when disturbed. It is also found on the sides of ditches, and in the autumn will seek cover amid the patches of potatoes and turnips.

GALLINULA CHLOROPUS, *Water-Hen.*—Single birds have occurred as stragglers on the following occasions—on 15th January 1906; 15th February 1908; 2nd May 1908; 6th May 1911; and 16th November 1908. It is a native bird, but not common, in Shetland.

FULICA ATRA, *Coot.*—Has occurred on a few occasions only. Single birds have been observed on 8th January 1909; 25th March 1909; 26th March 1908; 21st April

1909; 9th and 26th October 1908. From 4th to 9th May 1910, two were seen. The bird breeds as near as the southern portion of the mainland of Shetland.

ÆGIALITIS HIATICOLA, *Ringed Plover*.—One or two pairs are summer visitors to the Isle; but many appear regularly on passage in both spring and autumn. Single birds have arrived at the end of January, and one or two in February and the first half of March. After the latter date, numbers, sometimes many, appear down to the end of March, when the movements cease to be observed.

In autumn the returning birds arrive during the first week of September, and the movements are in progress until mid-October, and in some seasons mid-November. A single bird was seen on 30th November 1910, and another on 13th December 1910.

CHARADRIUS PLUVIALIS, *Golden Plover*.—A bird of double passage, being common at both seasons.

The spring movements northwards commence with the earliest days of March, and are observed throughout April and May. The migrants are most abundant between the latter half of April and the middle of May, but small numbers have been seen down to the end of the latter month.

In autumn a few appear in July and August, probably from Shetland; but it is not numerous until the latter half of September. The numbers fall off in October, and after the middle of the month, and during November and December, stragglers only have been seen.

SQUATAROLA HELVETICA, *Grey Plover*.—A single bird only has as yet been detected. This I saw on 23rd September, and again on the 25th, in 1908.

VANELLUS VANELLUS, *Lapwing*.—An abundant and regular visitor on passage during spring and autumn. Has rarely occurred in winter.

The spring migrants appear, as a rule, during the first week of March, and their movements northwards last until the end of the first week of May. On 21st May 1910, ten arrived, and stragglers have been observed as late as 17th June, in rushes of other species. It appears in flocks, often of considerable magnitude, some of which have been estimated to contain as many as six hundred birds or more.

A number, probably from Shetland, appear towards the end of June, and during July and August; but the more northern immigrants do not arrive until later, and not in numbers until the first week of October, and occur until the early days of November. Stragglers appear later and on one or two occasions in December. On 1st January 1909, strange to relate, a number appeared, over seventy being counted. There are other records for the first half of January, which would seem to indicate an exceptionally early return to nesting-haunts in Shetland.

The Lapwing formerly bred in fair numbers on the island, and a pair nested in 1898.

STREPSILAS INTERPRES, *Turnstone*.—A bird of double passage, a few also occasionally occurring in winter.

Has appeared as early as 3rd March, but the whole of May is the chief period for its passages in spring.

In autumn its visits date from 24th July, followed by a few in August, fair numbers in September, and a few again in October and during the first half of November.

This species is never very numerous, probably because

the Isle only affords very limited haunts—low rocks between tidemarks—suited to the habits of this species.

HÆMATOPUS OSTRALEGUS, *Oyster-catcher.*—Is mainly a bird of passage, but a considerable number are summer visitors to the Isle, and nest. One or two occasionally occur in winter.

Single Oyster-catchers have arrived as early as 10th February, but the main body of the native birds does not appear until the first half of March. As the summer visitors are numerous, it is difficult to discriminate between them and the transient visitors. The latest date on which autumn migrants have been seen is 12th October.

PHALAROPUS FULICARIUS, *Grey Phalarope.* — Two occurrences only have been detected of the visits of this species. Single birds appeared on two occasions in January 1909—namely, one on the 2nd, which was obtained, and another on the 29th. These are interesting records, since they seem to indicate that this species was wintering not very far off the Isle.

PHALAROPUS HYPERBOREUS, *Red-necked Phalarope.*—A male was captured in the south bay on 4th June 1910. This is the only known occurrence, though the bird is a summer visitor to other isles of the Shetland group.

SCOLOPAX RUSTICULA, *Woodcock.*—A regular visitor on passage at both seasons; but most abundant in the autumn, when it sometimes occurs in extraordinary numbers.

The spring movements have been known to commence during the latter days of March (24th earliest), and are observed throughout April and during the first week of May.

In autumn there are two records for September,

single birds having occurred on the 23rd and on the 30th. A few appear annually during the first half of October, but it is not until after the middle of the month and during the first week of November that considerable numbers arrive. There are usually two pronounced arrivals, which sometimes take the form of great rushes, each autumn: one in the latter half of October, and the other early in November. The latest date for stragglers is 22nd November. One or two, however, have appeared in December and January, their arrival being contemporaneous with an outbreak of frost and snow; such immigrants were probably birds that were wintering in Shetland.

During their sojourns on the Isle, which are of short duration, the birds frequent the high heathery ground and the hillsides, and are only occasionally found in the crofted area.

GALLINAGO MAJOR, *Great Snipe.*—There are four records of visits on passage of this species, and these relate to its appearance in spring and autumn. The bird is probably a regular visitor in small numbers at both seasons. The dates for its known occurrences are as follows:—5th May 1908, two; 15th May 1910, one; 5th September 1906, one; and 25th September 1909, one.

GALLINAGO GALLINAGO, *Common Snipe.*—Does not nest on the Isle, but a few winter. Many, however, occur during both seasons of passage.

The spring movements northwards have been witnessed between 9th April and 9th May.

A few appear in the latter half of July and during August, coming most probably from Shetland. Numbers arrive from the early days of September, throughout

October, and until mid-November. Stragglers occur later until 20th November. A spell of severe cold in December sometimes brings a few, sometimes many visitors—no doubt from further north, and most likely from Shetland.

GALLINAGO GALLINULA, *Jack Snipe*.—In spring this species has appeared in small numbers on its way northwards, from 24th March to 20th May.

On autumn passage it is frequent and abundant, arriving regularly in September (11th the earliest date) and throughout October. The records for its visits in November are not very numerous, and relate to small numbers. Single birds have been found in the first half of December, and there is a record of one on 16th January.

This species is frequently killed at the lanterns of the lighthouses—more so than any other Limicoline bird.

TRINGA ALPINA, *Dunlin*.—Is a bird of passage only, and common at both seasons.

It appears in spring during the second week of March, and its movements are in progress until near mid-June; but it is most abundant during May.

A few arrive annually towards the end of July. It comes abundantly late in August; occurs at intervals during September; but only occasionally appears in small numbers during October and the earliest days of November.

The small race—the "*Pelidna schinzii*" of Brehm—has twice come under my notice. On 12th September 1905, I shot two in widely separated parts of the Isle; and from 4th to 12th September in the following year, several were observed in small flocks, some of the members of which were still in partial summer plumage.

It is not improbable that this form may occur annually, but the haunts of waders are almost impenetrable (see p. 52).

TRINGA MINUTA, *Little Stint.*—Single birds have visited the Isle annually in the autumn, the dates of their appearance ranging from 14th August to 9th November. Most of the occurrences have been in September, and there is no instance of a visit in October. The November bird was sent to me. It is an adult female, still retains much of its summer plumage, and is in moult.

TRINGA TEMMINCKI, *Temminck's Stint.*—A bird of the year was captured on 14th August 1908, and sent for identification. This species had only previously been recorded for Scotland from the Aberdeenshire coast.

TRINGA MARITIMA, *Purple Sandpiper.*—A common winter visitor. Also occurs as a bird of passage.

The earliest date for its arrival in autumn is 19th August. It is fairly common in September, but not usually numerous until the latter half of October. In spring it has been observed as late as the first week in June.

During the winter gales, these birds are evicted from their haunts on the reefs, and seek the burn sides and even the high heathery ground.

On the 10th of May 1910, I surprised a party of Purple Sandpipers feeding on low rocks fringing the south bay. They launched themselves into the water without the slightest hesitation, and swam leisurely away, seeming to prefer this mode of retreat to taking wing.

TRINGA CANUTUS, *Knot.*—The rock-bound coasts of Fair Isle do not often tempt the Knot to seek them. A few, chiefly single birds, have occurred on the autumn passage, at dates ranging from the second week in August to 30th November.

Calidris arenaria, *Sanderling*.—A few visit the Isle on passage annually, when they are usually to be found on the little strip of sand which forms the head of the North Haven.

The spring visitors have been observed at dates ranging from 25th May to 11th June; while those in the autumn have appeared between 24th July and 27th September, but are most frequent during the latter month. One was obtained on 14th November 1911.

Machetes pugnax, *Ruff*.—Occurs on the autumn passage, but has not yet been detected in spring. The dates of its visits range from 24th August to 12th September; and four is the largest number of birds observed on any single day.

Totanus hypoleucus, *Common Sandpiper*.—Occurs in some numbers on both its spring and autumn passage movements.

In spring the visits on its way northwards have been chronicled between 23rd April and 16th June, but the birds are most frequent and numerous during the latter half of May.

The return autumn movements have usually commenced with September, and cease with the month. In 1911, however, odd birds appeared late in July (25th earliest) and during August, and were probably summer visitors leaving Shetland, where it nests in limited numbers. A few have occurred some years in October (as late as the 29th), and in 1908 one appeared with a rush of migrants on 2nd November.

On its passages this bird frequents the rocks and

reefs bordering the sea, and is seldom seen by the burn sides.

Totanus glareola, *Wood-Sandpiper.*—There is only one record for the occurrence of this species—namely, a male, on 26th May 1908, when a rush of migrants appeared (see p. 71).

Totanus ochropus, *Green Sandpiper.*—A few occur regularly on passage each autumn; but, as yet, we have only a few records of its visits in spring.

The autumn appearances date from 29th July to 14th September. As a rule, single birds are seen, rarely two, but on 11th August 1910, three were seen.

In spring, single birds were noted on 6th May 1908 and 16th May 1910, the last remaining until the 20th. In 1911 a single bird appeared on 6th May, and two on the 12th.

Totanus flavipes, *Yellowshank.*—A Yellowshank was shot on 24th September 1910, on the margin of one of the ponds which supplies the water for the primitive mills in which corn is ground. This bird is an exceedingly rare visitor to the British Isles, where it has only been known to occur on two previous occasions, and is new to Scotland.

The "Yellowlegs" is a native of America, where it summers and nests in Alaska, Canada, and Labrador; migrates through the United States, and winters in South America, proceeding as far south as Chili and Patagonia.

Totanus calidris, *Redshank.*—A common bird of passage. Has occurred in winter, but its appearances at that season seem to be quite exceptional.

Single birds have been seen in February, March, and early April; but it is only towards the end of the last-named month, during May, and the first week in June, that it occurs in some numbers.

In autumn it has appeared as early as 13th July, and as late as 2nd November, being most abundant and frequent during August and September.

TOTANUS NEBULARIUS, *Greenshank.*—Until the year 1910, there was only a single record of a solitary example, which appeared on 21st August 1908. In 1910, one was seen on 30th August, and on 1st September no fewer than ten were observed. The only spring record is of one seen on 15th April.

LIMOSA LAPPONICA, *Bar-tailed Godwit.*—I saw a bird of the year near the south lighthouse on 9th September 1908. This is the only known instance of the visit of this common northern species to the island, and affords significant testimony to the unsuitability of the Isle as a resort for migratory wading birds.

LIMOSA LIMOSA, *Black-tailed Godwit.*—A male was shot in a moist place amid the crofts on 8th January 1908. The occurrence of this species in Scotland in winter has hitherto been unknown, and to find it as far north as Fair Isle, at that season, is certainly remarkable.

NUMENIUS ARQUATA, *Curlew.*—Common on both seasonal passages, and of irregular occurrence during winter.

Migrants appear some years during the last few days of February, but early March is the ordinary date for the first visits of the season. The passages are in progress until late in May, and sometimes early June, and considerable numbers have been observed in rushes with other migrants.

A few appear in July, many in August, and varying

numbers throughout September and October and until the end of the third week of November.

NUMENIUS PHÆOPUS, *Whimbrel*.—An abundant and regular visitor on both its seasonal journeys.

In spring it is mainly seen on the land during its short sojourns, usually engaged in searching for beetles and other insects. In the autumn the reefs and rocks skirting the tide are its favourite resorts.

The earliest date for its appearance in spring is 15th April. The passages are in progress throughout May and until the end of the first week of June; but single birds, as stragglers, have been seen to near the end of the latter month.

In early August numbers appear, probably emigrants from Shetland, and the movements southwards are in progress during September. A few have occurred occasionally in October; and in 1907 and 1911, several arrived in November, one of which was seen in company with Curlews as late as 12th December 1907.

STERNA FLUVIATILIS, *Common Tern*.—The only record of the occurrence of this species is one of considerable interest.

At 10.30 on the night of 11th September 1906, the attention of Mr Kinnear and myself was drawn, by their loud cries, to a number of migrants which suddenly appeared at the lantern of the south lighthouse. These cries, we found, were uttered by a number of Terns, which were careering wildly around the light. We proceeded to the gallery, and succeeded in securing a number of the noisy visitors as they struck or fluttered against the windows in their vain endeavours to reach the light. These captures, we found, belonged to both the Common and the Arctic species. The birds remained flying in the rays and striking against the lantern until 2 A.M., when, owing to the appearance of fog, the dismal and terrifying blasts from the fog-horn were turned on. These soon scared the visitors, and saved them from the mystifying and dangerous influence of the brilliant beams of light. Herons, Wheatears, and White Wagtails were also present, the weather being favourable for the display of the decoying powers of the lantern. On the two following days a number of Terns, whose identity could not be determined, were present off the south end of the island—the only instance of the presence of these birds during the daytime that has come under my notice. On the night of the 20th, at 9 P.M., a party of Terns on migration again appeared in the vicinity of the lighthouse, but as the night was clear they did not approach the lantern, and only announced their presence by their cries. This, and much other evidence in my possession, proves that Terns largely migrate during the hours of darkness.

The Common Tern has only during recent years been added to the list of the breeding birds found in Shetland. The above is the only instance in which this species has been identified as a visitor to Fair Isle.

STERNA MACRURA, *Arctic Tern.*—Our introduction to this species as a Fair Isle bird has just been related under the Common Tern. On 22nd September 1909, a number noisily announced their appearance at 7.30 P.M. and flew in the rays of the lantern for a short time. On the following night a party arrived at 9.45 P.M., but did not remain long.

Terns on migration chiefly visit the Isle during the night-time and mostly pass unnoticed unless the weather conditions be such as to render the rays from the lantern conspicuous. No doubt these birds are to be seen at sea off the Isle, but they have only once come under my observation. Terns, probably this species, are said to have bred on the island in the early decades of the 19th century.

LARUS RIDIBUNDUS, *Black-headed Gull.*—Visits the Isle at intervals in winter, and is a common visitor before and after the nesting season.

It passes northwards in spring during March, April, and May, being most numerous during the latter month.

In autumn it occurs regularly in mid-July, and onwards to the end of September. A few are seen occasionally in October and November.

LARUS CANUS, *Common Gull.*—A few are seen in winter, but the bird is much in evidence during the seasons of passage. It has not been known to nest.

It appears late in April and during May, and a few are seen in the first week in June.

In autumn small numbers appear from the second week of July until mid-August, after which the bird becomes more numerous, and is observed until near the end of October. Very few have been seen later.

LARUS ARGENTATUS, *Herring-Gull.* — A common resident, nesting in considerable numbers. Many young birds desert the sea in the autumn, and endeavour to eke out an existence on the land, with the result that many perish. The young birds are much preyed upon by the Peregrine, and what this bird leaves of its victims often becomes the food of other Herring-Gulls, both old and young—hence the bird is guilty of cannibalism.

LARUS FUSCUS, *Lesser Black-backed Gull.*—This Gull is a common summer visitor to the Isle.

It makes its first appearance between 17th March and 5th April, and nests in numbers.

The main body departs in August; a few, chiefly young birds, are to be seen in September; and a single bird, an adult, has been observed as late as 22nd October.

LARUS MARINUS, *Great Black-backed Gull.*—This bird is numerous as a breeding species, and is present throughout the year.

LARUS GLAUCUS, *Glaucous Gull.*—Is a winter visitor, and occurs in some numbers. During its sojourn it is frequently seen on the land, like the Herring-Gull.

The first arrivals are chiefly young birds; and the earliest date for the appearance of the species is 5th October. There are no precise data relating to its departure, but it appears to leave early in the year. Unusual numbers appeared after the gales of 13th and 14th November 1910.

LARUS LEUCOPTERUS, *Iceland Gull.*—The recorded information regarding the visits of this species is far from voluminous, and relates to observations made in the months of November, December, January, and March. The earliest record is for 5th November, when a young male was captured and sent for identification, and the latest on 8th March, which was also sent.

RISSA TRIDACTYLA, *Kittiwake.*—A resident all the year round, passing the winter at sea off the Isle, and nesting in great numbers on the cliffs. This species has not been known to occur on the land in the roughest of weather.

MEGALESTRIS CATARRHACTES, *Great Skua.*—In 1905 the natives informed me, in reply to my enquiries, that the "Bonxie" formerly bred on the Isle. They were unable to fix the date, but the information had been handed down to them by their forbears. The accuracy of this traditional evidence is vouched for by Patrick Neill, who visited the northern isles in 1804. In his *Tour*, published in 1806, he informs us (p. 90, footnote) that this bird has "its principal breeding-place in the island of Faulah; but it breeds also in Fair Isle, and in one or two other places."

Odd birds, and as many as three in company, have been seen in early May and during June, probably visitors from the Foula colony. On 15th June 1909, one was seen on Fair Isle in hot pursuit of a Lapwing. In autumn it is observed in September and as late as 8th October.

STERCORARIUS POMATORHINUS, *Pomatorhine Skua.*—Doubtless of frequent occurrence at sea off the Isle, but has only twice been obtained and forwarded. Both these known visits have been in the autumn, and their dates are 29th October and 27th November.

STERCORARIUS CREPIDATUS, *Arctic Skua.*—This bird formerly nested in small numbers which were reduced to a single pair in 1905; since then it has not bred.

A few are seen at sea off the Isle in spring, summer, and autumn; but these appearances have little significance, since the bird is a common nesting species at Foula and elsewhere in Shetland.

ALCA TORDA, *Razorbill.*—An abundant breeding bird. All leave their summer haunts before the end of August and proceed to sea, where some pass the winter off the Isle. It returns to its rock nurseries in January, but does not finally take up residence on them until the last week of May.

ALCA IMPENNIS, *Great Auk.*—Messrs Baikie & Heddle, in their "Historia Naturalis Orcadensis," published in 1848, after stating (*Zoology*, Part I., p. 88), that this bird had not visited Orkney for many years, proceed to say that one was seen off Fair Isle in June 1798.

URIA TROILE, *Common Guillemot.*—The remarks on the Razorbill apply also to this species, except that during the autumn one or two are to be seen resting on the rocks close to the Isle. In early March a considerable increase takes place in the numbers of those observed at sea off the island, clearly indicating arrivals from afar. The cliffs are visited in January, and final residence for the summer is taken up late in May.

URIA GRYLLE, *Black Guillemot.*—A common resident and breeding bird. In the autumn and winter it seeks the shelter of the geos and bays.

ALLE ALLE, *Little Auk.*—Passes the winter in considerable numbers at sea just off the Isle. The earliest recorded date for its appearance is 22nd October; and the latest date for its stay in spring, 14th March.

FRATERCULA ARCTICA, *Puffin.*—Is practically a summer visitor, since it is rarely seen in winter at sea off the island. It appears late in March (21st earliest) and during early April, but does not finally take up its residence on *terra firma* until the early days of May. Many leave during the second week of August, and a little later all, save those still feeding young, have departed.

COLYMBUS GLACIALIS, *Great Northern Diver.*—Has only been detected on one or two occasions, in October and January. The Fair Isle seas, with their swift-running tidal races and lack of shelter, appear to be shunned by this bird.

COLYMBUS SEPTENTRIONALIS, *Red-throated Diver.*—The remarks under the last species apply equally to this.

The bird has only once come under notice—namely, on 10th September 1906.

Podicipes auritus, *Slavonian Grebe.*—A few appear regularly in the north and south bays in the autumn. Here it has been observed as early as 9th September, but the first week of October is the accustomed date for arrival, and its passages or sojourns extend to 28th November, which is the latest date on which it has been recorded as seen.

Podicipes fluviatilis, *Little Grebe.*—Has only come under observation on a few occasions.

In spring it has once appeared in a rush of migrants—namely, on 9th May 1908. It has been seen several times in the autumn between 12th and 27th October, and there is a single record of its occurrence in winter—on 9th December.

Procellaria pelagica, *Storm Petrel.*—Quite a number breed on the Isle. I have seen young birds in their rugged nurseries in September, and others have been found clad largely in down, as late as 31st October. On 2nd October 1907, one just able to fly was captured at the lantern of the south lighthouse. This juvenile was fully feathered on its back and neck, but its wings and tail had not nearly attained to full length, whilst its chest and abdomen were entirely a mass of down. How such a youngster could have reached the lofty beacon on such diminutive pinions is a marvel; and should it have ventured to alight on the water its nether plumage must have become as water-logged as a sponge. Old birds have been obtained on the island as late as 10th November.

Puffinus anglorum, *Manx Shearwater.*—Is observed off the Isle in both spring and autumn. In the former season its visits have been recorded between 25th May and 26th June; and at the latter from 25th September to 3rd October. No doubt many escape notice.

FULMARUS GLACIALIS, *Fulmar Petrel.*—Until the first or second years of the present century this bird was only occasionally observed; now it is an abundant native species. In 1902 it was first detected as being present all through the summer season, but no eggs were found. Since then it has increased immensely, and breeds in suitable places all round the island.

It occurs all the year round, but is only occasionally seen in October, November, and December. During the mild season of 1910 to 1911 it was common inshore all winter. It leaves its nesting-haunts early in September; appears again in mid-December or early in January; and takes up its breeding-haunts on the cliffs in earnest during the first week of March.

ADDENDA

EMBERIZA LEUCOCEPHALA, *Pine-Bunting.*—A male of this native of Siberia was obtained on 30th October 1911. It is not known to have previously visited the British Isles.

ACROCEPHALUS PALUSTRIS, *Marsh-Warbler.*—See p. (and footnote) 134.

ANAS STREPERA, *Gadwall.*—A young male, observed on 11th October 1911, is new to the avifauna of the island.

[*Photo: J. R. Russell.*

ST KILDA: THE VILLAGE AND BAY FROM THE ISLAND OF DUN.

CHAPTER XXII

BIRD-MIGRATION AT ST KILDA, BRITAIN'S OUTERMOST ISLE IN THE WESTERN OCEAN

ST KILDA enjoys the distinction of being the remotest of all the Isles of the British Seas. It lies—

> "Far in the watery waste, where his broad wave
> From world to world the vast Atlantic rolls"

—in latitude 57° 48′ 35″ N., and in longitude 8° 35′ 30″ W., and is over 40 miles west of the North Uist, the nearest island of the main outer Hebridean group.

From the year 1697, when Martin Martin published the quaint and engaging account of his visit to St Kilda, down to the present day, this tiny Atlantic archipelago has been surrounded by a halo of romantic interest which is still happily in the ascendant. It owes this unique place among our isles to a number of exceptional peculiarities and associations. Among these are: its extreme remoteness, already alluded to, as the home of the most isolated and hence unsophisticated community of Britons; the surpassing grandeur of its cliffs, stacks, and rock-scenery generally; and the marvellous hosts of sea-fowl which annually seek its fastnesses as a cradle and nursery wherein to rear their offspring. The further facts that it was formerly the home of the flightless Great Auk, is still the British metropolis of the Fulmar and Fork-tailed Petrels, and possesses a Wren and two species of Mice which are entirely peculiar to the group, have combined to make St Kilda famous in the annals of natural history.

St Kilda, too, has gained considerable fame as having been for centuries the home of a race of daring cragsmen who stand unrivalled in their dangerous calling. All the men follow this "dreadful trade," for by them the myriads of sea-fowl and their eggs have ever been considered of greater import than the rich harvest of fish which lies almost at their thresholds, but is little garnered.

Martin's remarkable, and on the whole singularly accurate account of St Kilda and its bird-life at the close of the seventeenth century, attracted much attention; and its interest was supplemented by another old-time observer, Macaulay, who visited the islands in 1758. Later, especially during recent years, the isles have become happy hunting-grounds for the collector in search of spoils wherewith to enrich his egg-cabinet, and this has led to the summer aspect of their bird-life being as well known as that of any portion of the British area, owing to the many descriptions of it that have been published.

Down to the autumn of 1910, the St Kilda birds had only been studied during the summer time, and it remained to be ascertained by what feathered visitors the isles were sought at other seasons, and especially during the periods of the great migratory movements. Here, then, was an interesting little problem in bird-migration awaiting solution. It was most desirable that

PLATE XVII.

[Photo: George Stout.

ST KILDA: THE HEAD OF THE BAY, THE VILLAGE, AND CONNACHER.

the investigations be undertaken, even if negative results alone rewarded the adventure, in order to ascertain what migrants, if any, visited this far-outlying and remotest of the British Isles. I had made arrangements to spend my autumn vacation there in 1904, but circumstances arose which compelled me to change my plans, and it was not until last year, 1910, that I again turned my thoughts towards that remote western archipelago.

To render such a visit possible in the autumn, the season best suited for the investigations, necessitated making private arrangements for my return to the mainland, for there is no regular communication with St Kilda from the end of August until May of the following year. Fortunately, through my friend Prof. D'Arcy Thompson, I was able to enlist the aid of the Scottish Fishery Board, who most kindly arranged that I should be relieved by one of their cruisers—an obligation which I gratefully acknowledge, though, as the sequel will show, I availed myself, under the pressure of circumstances, of other means of escape which unexpectedly presented themselves.

Leaving Glasgow on the afternoon of 29th August, accompanied by George Stout, lately my trusty observer at Fair Isle, I landed at St Kilda on the morning of 1st September, and took up residence in the factor's empty house, which had been most kindly placed at my disposal. Here we remained until 8th October, when the weather, which had been unsettled for some days, showed signs of growing worse, and gave warning that it was high time to emigrate, or to risk remaining on the island for a longer period than my leave warranted my doing. On the day named, I accepted the most kindly offer made to me by Capt. Donald Craig, of the Aberdeen trawler *Wamba*, of a passage to the island of Lewis.

Our second voyage to St Kilda almost resulted in the wreck of our hopes of reaching the island. We left Glasgow on 28th August 1911 by the last steamer of the year, and all went well until we emerged from the Sound of Harris on the early morning of the 31st. Here we found the Atlantic in a state of great unrest, due to the prevalence of a south-westerly gale, and after proceeding some twenty miles in the direction of St Kilda, it was found impossible to proceed further, owing to the increasing violence of the wind and waves. The S.S. *Hebrides* was compelled to return to Lochmaddy, where we left her. Our luck, however, had not entirely deserted us, for on our proceeding to Loch Tarbat on the following day, my friend Mr Carl Herlofson most kindly sent us out in one of his whaling steamers, and, after a rough passage, we slept at St Kilda only twenty-four hours later than we had originally hoped to do.

Along with George Stout, I remained six weeks on the island, and left on 12th October on board the steam trawler *Mercury*, of Hull, whose captain, William Rylatt, most obligingly conveyed us to Stornoway, and showed us great kindness and attention on the voyage. That the fishery cruiser again failed to relieve me, was in no way her fault; she was prevented by unexpected official work from visiting the island at the time arranged, but arrived there to take me off the day after I had flown. I have to thank the Fishery Board for the great obligation under which they again laid me: without their

[*Photo: George Stout.*

ST KILDA: THE MANSE, AND LOOKING WEST.

[*Photo: George Stout.*

ST KILDA: THE VILLAGE STREET.

valuable co-operation I should not have ventured to St Kilda so late in the season.

Quite a number of descriptions have been published of the five islands and the several stacks which form the St Kilda archipelago. Such being the case, it is only necessary to give a short account of the topography of the main island, Hirta, the scene of the investigations. Hirta is some two and a half square miles in area, and is an island of mountains and high ground. It falls naturally into two halves or basins, an eastern and a western, each of which is drained by a moderately sized burn. The low ground in both areas is of very limited extent.

The eastern section of the island is the site of the village and the cultivated ground, and has for its sea front a fine bay, which affords the only suitable landing-place in the island. This lowland area is most effectually sheltered from the east, north, and west, by a picturesque series of hills—namely, Oisaval (930 feet), Connacher (1372 feet), Mullach Mor (1153 feet), and Mullach Sgail (705 feet), which are connected by high ridges ranging from 540 to 850 feet. The village and the crofts are thus flanked on all the landward sides by high ground, while the long narrow island of Dun shields them from the south. The crofts, amid which the village stands, are surrounded by a stone wall, the object of which is to keep out the cattle and sheep which feed on the slopes beyond. The area thus enclosed is probably about 100 acres, but the actual amount under cultivation does not exceed some 50 acres. The sides of the hills surrounding the crofts rise rapidly, and, when free from masses of rough boulders and great screes, are clothed with short grass, intermixed with which is a little thin heather. Crags and rocky faces present themselves on the higher parts of Connacher and Mullach Sgail. Beyond, to the north and east, the summit rims of Oisaval, Connacher, and their connecting ridges present to the Atlantic a line of magnificent cliffs, the finest in the British Isles, ranging from 500 to 1262 feet in height. These are the summer homes of many thousands of Fulmars, and innumerable other rock-fowl—Guillemots, Razorbills, Puffins, and Gulls—all of which, except the Fulmars, had departed from these aerial nurseries, along with their young, ere we arrived.

The village, consisting of seventeen inhabited houses, with the crofts back and front of it, lies between the head of the bay and the foot of the hills, and its houses front at intervals a curved footway or narrow street. The beach is a massive rampart formed of sea-worn boulders, at the foot of which, at low tide, a strip of silver sand is exposed about its centre; this was much sought by the few wading birds that came under notice.

The western half of the island consists of a valley about three-quarters of a mile in length, and known as "The Glen." This is drained by a burn which discharges itself, after a northerly course, into a wide bay on the north-west side of the island. Like the low land of the eastern half, it is surrounded on three sides by hills, the highest of which are over 1100 feet on the east and south, and range from 1000 to 500 feet on the west. This high ground also presents to the Atlantic lofty, rugged cliffs and steep, grassy slopes capped with naked rock, except at the mouth of the Glen, where the coast is low and rocky.

The cultivated ground around the village, with its

PLATE XIX

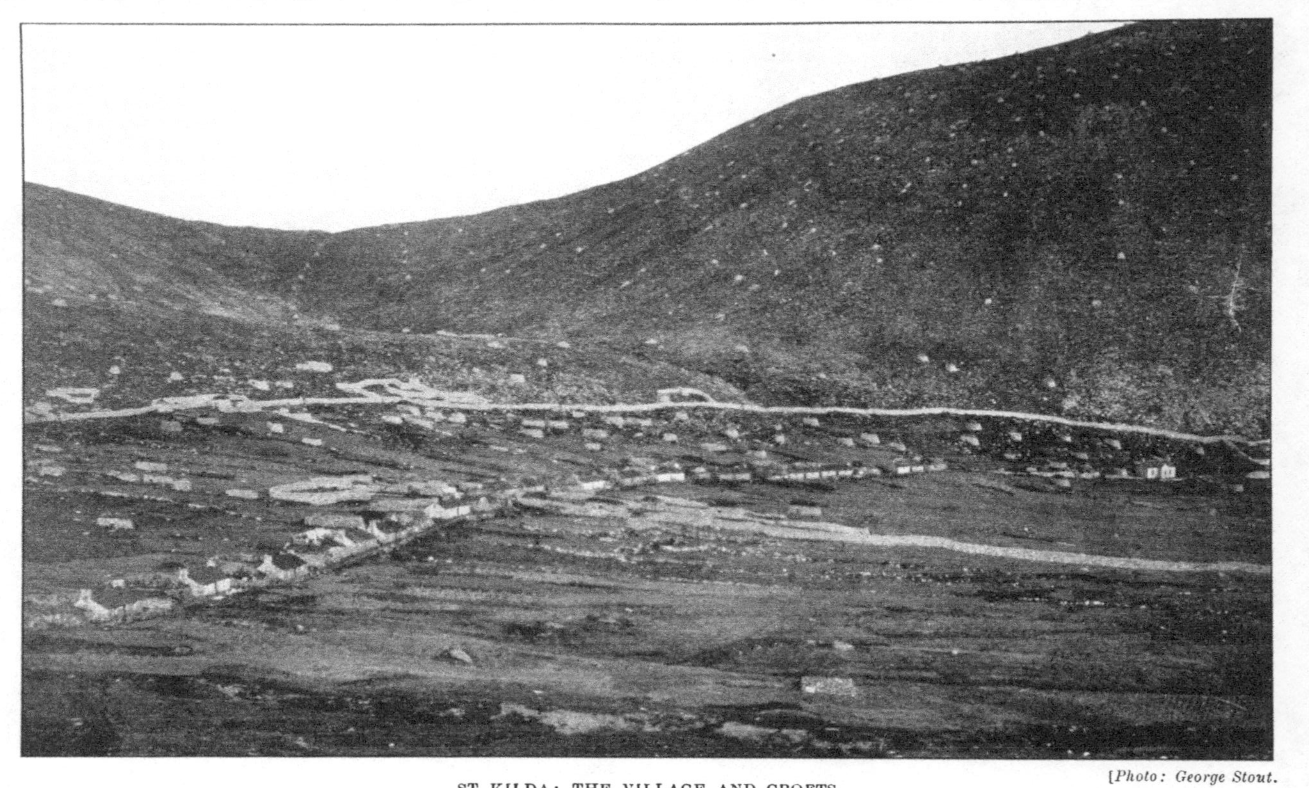

[*Photo: George Stout.*

ST KILDA: THE VILLAGE AND CROFTS.
The white spots are the stone erections called "cleits.

strips of oats, potatoes, and meadow land, here and there interspersed with patches of brackens, docks, nettles, and yellow iris, afford the only cover to be found in the island. There are no shrubs or trees of any kind. Thus the crofts were our main observing-grounds, and proved to be the resort of almost all the smaller migrants visiting the island. Certain outlying enclosures, in which grass was grown, the Manse garden, and the higher ground also received attention. The Glen, which looked so promising on Mr Heathcote's map, and is usually described as being fertile, proved most disappointing. There was no cover, and the ground was everywhere wet, and clad only in short grass, amid which much sphagnum flourished: several visits were paid to this locality, and it was on each occasion found to be singularly destitute of bird-life.

No account of St Kilda can be deemed complete which makes no mention of the vast number—many hundreds—of small stone erections in which the inhabitants keep their turf-fuel and hay, and formerly kept their stores of dried birds. These are to be seen everywhere, even on the faces of the cliffs and the highest ground. Martin alludes to their numbers and calls them pyramids—the modern St Kildan name is Cleits; many of those now existing were doubtless in use when Martin visited the island in 1697.

The high ground, some of which is table-land, yielded only a few species, such as Snow-Bunting, Lapland Bunting, Wheatears, Meadow-Pipits, Golden Plover, and Snipe.

The height of the cliffs was so stupendous, and their faces so precipitous, that it was quite impossible to ascertain whether they harboured migratory birds or not.

Those on the north, which we visited most, were tenanted for the greater part of our stays by vast numbers of Fulmars; and it is doubtful under these circumstances whether they would afford an agreeable resting-place for migrants.

Turning now to the results obtained through the investigations, these, it may at once be stated, greatly exceeded my most sanguine expectations; for I was quite prepared to find evidence of a negative nature at such a far-western and remote station. I saw much migration there on both my visits, though the weather conditions during the whole of September and the early days of October in 1910 and 1911 were distinctly unfavourable for great movements between the British Isles and the continent of Europe, and none were observed either at St Kilda or elsewhere. The types of weather alluded to resulted in winds persistently ranging from north to south *by west*. They were easterly, but unsettled, on one or two occasions only. The result of these weather influences was that the migrants moved southwards towards their winter quarters in dribblets and largely escaped notice, as they always do under such circumstances. The persistency of this westerly type of weather was most tantalising, and it is surprising that under its baneful influence so many migrants from the Continent came under notice. In 1911 the conditions were particularly unfavourable for the visits of migrants coming from the eastwards. Fortunately the weather changed during the early days of October, and then many of these feathered visitors rewarded our researches. This westerly type of weather, however, does not interfere with the continuous arrival of such species as are derived from

Iceland and Greenland, since it does not ordinarily interpose meteorological barriers between the north-west and St Kilda.

The investigations carried out at St Kilda in the autumns of 1910 and 1911 prove that the island is annually visited by considerable numbers of migratory birds.

It is an important station on a main route traversed by summer birds proceeding to and from Iceland, and by the few Palæarctic species which are migrant-natives of Greenland. This was abundantly evident during both visits from the numbers and the frequency of the arrivals of the Greater Wheatear, and hardly less so from those of the White Wagtail, Meadow-Pipit, Redwing, Snipe, and Golden Plover; and further, in 1911, from the occurrence of several Greater Redpolls, whose native home is in Greenland and North-Eastern America. The north-western element in the ornis of the archipelago is further strengthened by the discovery of the Pennsylvanian Pipit and Baird's Sandpiper—both of which are natives of North America.

St Kilda also lies on the very verge of that enormous stream of migration which rushes along the British coasts to and from Northern Europe, and in these remote islands reaches its extreme westerly limit in northern Britain. During my visits I found no less than twenty-six different kinds of such European birds —species which change their homes with the seasons— as migrants on their southern journey to accustomed winter quarters.

In all, ninety-six species came under our notice. Of these, sixty-two were migrant-visitors chiefly on their passage southwards to winter quarters, and fourteen were mere waifs. No less than forty-eight of these species were new to the avifauna of the St Kilda group; and not a few had not previously been observed in any of the isles of the Outer Hebrides. The rest of the birds noticed were natives of the archipelago, some of which are more or less migratory, since they seek the islands in spring and depart for a longer or shorter time in the autumn: among these are the Gannet, Guillemot, Razorbill, Puffin, and certain Gulls and Petrels.

Among the species observed were several of considerable rarity, including one new to the British avifauna and two to that of Scotland. These came as pleasant surprises, but, as in similar cases, not always — far from it — as aids to the solution of migration problems. The new bird for Britain was the American Pipit (*Anthus pennsylvanicus*), and those new to Scotland were the Marsh-Warbler (*Acrocephalus palustris*) and Baird's Sandpiper (*Tringa bairdi*). The other rare or little-observed species were several Scarlet Grosbeaks, Greater Redpolls, Ortolan Buntings, Little Buntings, Lapland Buntings, a Crossbill, Barred Warbler, Black-throated Chat, two Red-throated Pipits, Turtle-Dove, Ruff, Great Snipe, and a number of Sooty Shearwaters.

Other interesting visitors were the Chaffinch, Greenfinch, Whinchat, Redstart, Garden-Warbler, Whitethroat, Lesser Whitethroat, Blackcap, Willow-Warbler, Chiffchaff, Tree-Pipit, Brambling, Wryneck, and Coot. It is of extreme interest to have discovered that St Kilda is visited by such a number of birds which pass the summer in Northern Continental Europe.

The most abundant species observed on passage

were the Greater Wheatear, White Wagtail, and Meadow-Pipit, all of which are common summer visitors to Iceland, and hence, on the geographical grounds already mentioned, their appearance in numbers at St Kilda is not surprising.

Full particulars of the occurrences of the birds named and of the rest of the species observed by myself and previous visitors to St Kilda will be found in the succeeding chapter.

As I did not land on Soay, Boreray, or Levenish, or on the very remarkable stacks of Lii and An Armin—all of which are the resort of vast numbers of sea-fowl during the summer—I would refer anyone desiring information regarding them to Mr Heathcote's *St Kilda*, or to Dr Wiglesworth's *St Kilda and its Birds*. On the 8th of October 1910, we steamed close under the great cliffs of Boreray, in an interval when the island was brilliantly illumined by a gleam of sunshine which burst through the mist after a rain squall. The scene then presented was singularly beautiful and impressive—a study in dark frowning cliffs, slashed with steep grassy slopes of the brightest emerald green. Though only half a square mile in area, Boreray rises to the majestic height of 1200 feet, is everywhere precipitous, and is further remarkable for the rugged grandeur of its contours. From a scenic point of view, I should doubt if it has a rival in British seas. Gannets still crowded its ledges, and some of the young were being tossed in the cradle of the deep, apparently in a helpless condition.

Puffins, Guillemots, Razorbills, and Fulmars were still in the vicinity of the island. Of greater interest was the presence of numerous Great Shearwaters, as many as fifty of these active birds being seen dashing about on the wing simultaneously. A Sooty Shearwater and a Great Skua were also present.

I found all the resident small birds, except the Tree-Sparrow, remarkably tame and confiding, owing to the entire absence of all forms of persecution. Grey Crows sat upon the roofs of the houses, or fed on garbage at their thresholds; Starlings allowed one to approach them to almost within arms-length; and Rock-Pipits and Wrens entered the house in which I resided by both the door and windows. The fact that the Tree-Sparrow is as wild here as elsewhere, proves that extreme wariness and suspicion are deep-seated traits in its nature.

In giving this account of my experiences at St Kilda, I deem it a duty to say a few words about the inhabitants. Much has been written concerning them, and not all that has been said is of a flattering nature. During the eleven and a half weeks I resided among them, most of my time was spent in their crofts, and hence in their very midst, and I thus had unusual opportunities of forming an accurate opinion of their characters and habits. I found them exceedingly polite on all occasions, extremely kind and good-natured, much interested in my work, and most wishful to assist me in every way. They were all very industrious, and to this must be attributed the fact that there is no poverty among them. They are said to be greedy. I certainly did not find them so. I had to seek their assistance in connection with the landing and embarkation of my baggage and stores, and I had other dealings with them, and the remuneration asked was on all occasions most reasonable. I have every reason to think well of them, and I trust that I have not paid my last visit to their hospitable and most interesting island.

Diary of Observations, 1910.

The following Diary gives the names of the migratory birds observed during each day; also the weather conditions prevailing. The significance of the latter will be made evident on consulting Chapter VIII. :—

1st September.—North-west, fresh; bright sunshine.

The following migrants came under notice:—Barred Warbler, one; Willow-Warbler, several; Tree-Pipit, four or five; Ortolan Bunting, several; Common Wheatear, a few; Pied Wagtail, remains of freshly killed male found; White Wagtail, numerous; Meadow-Pipit, common; Ringed Plover, two; Turnstone, five; Sanderling, one; Lesser Black-backed Gull, two birds of the year.

2nd September.—North-west, strong breeze; migrants only found in sheltered places.

Garden-Warbler, one; Willow-Warbler, one; White Wagtail, twenty to thirty in party and many scattered birds; Tree-Pipit, two; Snipe, one; Dunlin, one; Merlin, one; Sanderling, three; Golden Plover, five; Ringed Plover, seven.

3rd September.—West, strong breeze; dull; warm.

Greater Wheatear, three; Ortolan Bunting, one; Willow-Warbler, two; Golden Plover, numbers; Garden-Warbler, two; Ruff, male; Curlew, one.

4th September.—South, light winds; warm; sunny at times.

Curlew, three; Ruff still here; Black Guillemot, one;

Willow-Warbler, one; Greater Wheatear, several; Ringed Plover, six; Heron, one at night.

5*th September*.—South, light; fog; sun at intervals in forenoon. West; heavy rain.

Whinchat, one; Tree-Pipit, one; Ruff still here; Willow-Warbler, three; Greater Wheatear, several; Wheatear, two; Curlew, twenty in party; Heron, one; Merganser, female.

6*th September*.—North-west, light; beautiful day. A number of migrants in.

Garden-Warbler, several; Willow-Warbler, several; Ruff still here; Marsh-Warbler, one; Common White-throat, one; Chiffchaff, several; Great Snipe, one; White Wagtail, several; Wheatear, a number; Tree-Pipit, one; Curlew, several; Wryneck, one.

7*th September*.—South-east to south-west, very light; bright; hot; sunny. Very few birds; yesterday's migrants have passed on.

Willow-Warbler, one; Garden-Warbler, one; White Wagtail, one or two; Curlew, one or two; Ringed Plover, one or two.

8*th September*.—South-west breeze; sunny.

Scarlet Grosbeak, two; Greater Wheatear, several; White Wagtail, two or three; Ringed Plover, one; Curlew, several; Oyster-catcher, one; Kestrel, one.

9*th September*.—South-west; dull; rain in afternoon and at night.

Willow-Warbler, one; White Wagtail, three; Curlew, four; Greater Wheatear, several; Turnstone, two; Ringed Plover, two.

10*th September*.—South-south-west in morning, with heavy rain; north later, with heavy drizzle.

White Wagtail, several; Ringed Plover, party; Golden Plover, one; Curlew, two; Knot, one.

11*th September.*—North-east; mist and drizzle until mid-day; then clear, with sunny periods.

Ringed Plover, about twenty in flock; White Wagtail, about a dozen; Garden-Warbler, one; Black Guillemot in the bay; Wheatear, numerous; Turnstone heard; Curlew, several; Greater Wheatear, several.

12*th September.*—Variable light airs, with drizzle and sunshine at intervals. Migrants more numerous.

White Wagtail, numerous; Greater Wheatear, numerous; Redstart, one; Common Snipe swarming among grass; Heron, two or more at night; Sooty Shearwater at sea off north-east of island.

13*th September.*—North, light; fine day; more migrants in.

Willow-Warbler, one or two; Ruff still here; Teal, male; Lesser Whitethroat, one; Turtle-Dove, one; Greater Wheatear, many; Meadow-Pipit, many; Golden Plover, four.

14*th September.*—North, light; very fine.

Lesser Whitethroat, several; Pintail, male; Merlin, one; White Wagtail, many; Meadow-Pipit, many; Ringed Plover, twelve; Herons again, at night.

15*th September.*—North-west, light; drizzle in morning; sunny in afternoon.

Lesser Whitethroat, one; Snipe, one; Ringed Plover, usual number seen; Greater Wheatear, a number; Curlew, four; Meadow-Pipit, many; Turnstone, one; Herons at night, as usual.

16*th September.* — North-west; dull and sunny periods; some drizzle; fine later.

Greater Wheatear, several; Ortolan, one; Snipe,

one; Merlin, one; Meadow-Pipit, many; White Wagtail, a few.

17*th September.*—West, light; dull and sunny periods.

Lesser Whitethroat, one; White Wagtail, several; Meadow-Pipit, very numerous; Crossbill, one; Greater Wheatear, a number; Curlew, several; Merlin, one.

18*th September.*—South-west breeze, after rain at night; dull and sunny periods. Heavy rains in afternoon and at night, with wind north-west.

Ringed-Plover, two; White Wagtail, two; Greater Wheatear, several; Meadow-Pipit, many; Curlew, several.

19*th September.*—North-north-east, strong breeze; cold; showers at intervals in morning; sunny later.

Merlin, two; Wheatear, about six; Meadow-Pipits, many; Willow-Warbler, one; Scarlet Grosbeak, female; White Wagtail, several; Garden-Warbler, one; Tree-Pipit, one; Kestrel, one.

20*th September.*—North-west; cool at first, then sunny and warm.

White Wagtail, three; Greater Wheatear, a few; Willow-Warbler, two; Redwing, one; Crossbill, female; Ringed Plover, four; Curlew, one; Meadow-Pipit, plentiful; Snipe, one; Lapland Bunting, one or two.

21*st September.*—West breeze, gusty; sunny.

Garden-Warbler, one; Teal, male; Ring Plover, four; Tree-Pipit, one; Lapland Bunting, one; Red-throated Pipit, one; Meadow-Pipit, still very many; Wheatear, a few; White Wagtail, two; Curlew, one; Snipe, one.

22*nd September.*—West-south-west, strong breeze, very gusty; dull but mild.

Greater Wheatear, a few; Curlew, one; Merlin, one; Meadow-Pipit, very numerous; Merganser, one; Ringed Plover, one; White Wagtail, absent.

23rd September.—South-west, strong breeze; very gusty; dull; rain at intervals.

Curlew, three; Turnstone, one; Ringed Plover, one; Greater Wheatear, several; Lapland Bunting, one; White Wagtail, one; Meadow-Pipit, many.

24th September.—West to south-east, light, fine; sunny; heavy rain at night.

Lapland Bunting, one; Coot, one; Heron, one; Greater Wheatear, two only; Ringed Plover, one; Turnstone, one; Meadow-Pipit, many.

25th September.—South-south-east breeze, drizzle at first, then fine; fog at night.

Meadow-Pipits, the only migrants seen.

26th September.—South-east, light, dull, mild; rain in afternoon and at night.

Redstart, two; Willow-Warbler, one; Merlin, two; Meadow-Pipit, many; Greater Wheatear, about a dozen: Heron, one; Snipe, one; Curlew, two.

27th September.—South-east to South-west; heavy rain until evening.

Greater Wheatear, six or more; Meadow-Pipit, many; Snipe, one; Golden Plover, one.

28th September.—South-west breeze; sunny; showers at intervals.

Curlew, one or two; Snipe, two; Ringed Plover, two; Greater Wheatear, several; Meadow-Pipit, many; Wheatear, three or more seen; Merlin, one; Golden Plover, eleven; Snow Bunting, one; Grey Plover, one.

29th September.—West, strong; sunny; heavy showers at intervals.

Meadow-Pipit, very abundant; Greater Wheatear, three; Golden Plover, five or six; Ringed Plover, two; Curlew heard; Snipe, one.

30th September.—North-west, west, and south-west breezes; fine. South-west, rising at night.

Meadow-Pipit, still many; Greater Wheatear, one; Heron, one; Ringed Plover, two; Golden Plover heard; Merlin, on croft; Pennsylvanian Pipit, one.

1st October.—South-west, high wind; heavy rain all day; mild.

Meadow-Pipit, many; Greater Wheatear, one; Grey Wagtail, adult male.

2nd October.—South-east gale; continuous heavy rain in morning. North-west, light; showery in afternoon.

Great Northern Diver in bay; Heron at night.

3rd October.—North breeze, gusty; showers and sunshine. At night, north-west, heavy drizzle.

Brambling, two or three; Wheatear, three; Grey Wagtail, one; Meadow-Pipit, many; Ringed Plover, one; Peregrine, three; Merlin, one.

4th October.—West, strong breeze; heavy drizzle; very mild.

Wheatear, two; Meadow-Pipit, many still; Grey Wagtail, female; Merlin, one; Curlew, one.

5th October.—West breeze, mild; heavy drizzle in morning and at night.

Greater Wheatear, one; Merganser, one in bay; Meadow-Pipit, numerous.

6th October.—South-west breeze; dull; mild.

Greater Wheatear, three; Tree-Pipit, one; Meadow-Pipit, many; Merlin, one; Blackbird, young male; Grey Wagtail, one; Curlew heard.

7th October.—South-west, light; a day of continuous pouring rain.

8th October.—North-west, fresh breeze. Left St Kilda at 11 A.M. for Braes-cleit, Isle of Lewis.

Great Shearwater, many off island; Sooty Shearwater, several off Boreray; Great Skua, one off Boreray and others seen; Richardson's Skua off Boreray and at intervals on voyage; Pomatorhine Skua, several seen.

In 1911, the weather from the 1st September to the 3rd October had been persistently of the westerly type and unfavourable for the appearance of migrants from continental Europe. On the latter date it changed, and easterly breezes and calms prevailed down to the 12th, my last day at St Kilda.

During this favourable period, immigrants from the north-east appeared, some of them in numbers, and included Chaffinches, Bramblings, Greenfinches, Skylarks, Redwings, Fieldfares, Blackcap, Lesser Whitethroat, Willow-Warbler, Chiffchaff, and Woodcocks. In addition, Mealy Redpolls, Mallard, Wigeon, Pintail, Long-tailed Ducks, Golden Plovers, Snipe, and Slavonian Grebe were also among the late arrivals. Some of these had been previously observed, but as their summer range includes Iceland, they may have been arrivals from the north-west.

CHAPTER XXIII

THE BIRDS OF ST KILDA, WITH SPECIAL REFERENCE TO THE MIGRATORY VISITORS

As a necessary preliminary to my proposed investigations, I made an examination of the somewhat extensive literature devoted to St Kilda, the object of this research being to ascertain what was already known regarding its birds. Such of this historical matter as bears upon their migrations has been incorporated under the few species to which it relates. In addition, it has been thought desirable to include within the scope of this contribution a concise historical account of all the species, both natives and visitants, which have been known to occur in the St Kilda group down to the end of the year 1911.

A list of the various works which have contributed more or less to our knowledge of the avifauna of the archipelago will be found enumerated, in their chronological sequence, in the Bibliography which forms the concluding portion of this Chapter. A number of other books, it is true, have been published on St Kilda, and contain references to its birds. Some of these, such as Buchanan (1793) and Macdonald (1841) have obviously borrowed all their references relating to the ornithology of the island from previous writers, while other authors have written, pleasantly enough, about their personal experiences, but have added little or nothing to our knowledge. Such works it has not been thought expedient to include in the list, but allusion has been made to some of them when necessary.

Of the older writers, Martin and Macaulay stand pre-eminent, and their information afforded the basis of all that followed until the year 1831, when Atkinson ushered in the modern era in St Kildan ornithology.

It should be remarked, however, that previous to the works of these authors, Sir George Mackenzie, the Lord Clerk Register, furnished a short account of Hirta to Sir Robert Sibbald between the years 1681 and 1685 (Mackenzie was created Viscount Tarbat in the latter year), in which he alluded to the Garefowl, and to the incredible number of fowls which frequent and cover the rocks, and when they rise darken the sky. From what source Sir George obtained this information is not known; and it is not until Martin's visit to the island in 1697 that we have any first-hand information regarding the bird-life of St Kilda.

Since my first visit, I have had most obligingly placed at my disposal by Dr Wiglesworth, for incorporation in this work, a series of notes on the occurrence of certain birds sent to him by Neil Ferguson since the publication of his excellent little book. These will be found under the species to which they refer.

As the result of these searchings among St Kildan literature, I found that seventy-three species of birds had been recorded as natives and visitors to the islands. To these I am able to add no fewer than forty-seven species hitherto unknown; forty-four through direct personal observation, and the rest from reliable information which will be duly acknowledged under the species to which it relates.

When the claims of any bird have been considered open to doubt, the species has been enclosed within square brackets. I wish it to be understood, however, that this merely indicates my personal opinion on its claims to be admitted as a member of the St Kilda avifauna.

The species marked with an asterisk are new to the avifauna.

CORVUS CORAX, *Raven.*—This species was first alluded to by Macaulay (p. 160), who visited St Kilda in 1758, and who states that at Hirta alone there were some Ravens of the largest size. In 1840 MacGillivray describes it (p. 160) as found in some numbers, and the same was said of it in 1841 by Wilson. Dixon, in 1885, remarks that Mr Mackenzie, the factor, saw seven pairs there at one time, but these may have included young and old.

I saw Ravens frequently, sometimes as many as six on the wing simultaneously, but was assured that now only two pairs are natives of the archipelago.

CORVUS CORNIX, *Grey Crow.* — Martin (p. 46) mentions "Crows" as seen by him in 1697, and they are alluded to by almost all subsequent writers. MacGillivray (p. 57), in 1840, says that he frequently saw a dozen on the roofs of one of the huts.

This bird is a resident and far too numerous, as many as forty being seen on the wing at once. It is also audaciously tame. They seemed noisily to resent our presence, and we shot a number of them to oblige the natives, who complained of the damage done to their crops, and destruction to their small stock of poultry.

CORVUS FRUGILEGUS, *Rook.*—Wilson (pp. 74 and 81)

states that the Rev. Neil Mackenzie informed him that this bird was said to have nested on the island in January, 1841, and that it occasionally visits the island in winter. The Rev. Mr Fiddes told Steele Elliot (p. 284) that hundreds passed over the islands in the winter of 1893, and that great numbers perished. Five were seen by Mr Steele Elliot in June 1894; the remnant of the above mentioned flock.

[PICA PICA, *Magpie.*—Macaulay (p. 160) tells us that Magpies "have been seen in the isle more than once, though very seldom in any other part of the western Æbuda."]

STURNUS VULGARIS, *Starling.*—Starlings are mentioned by Macaulay (p. 160), who visited the island in 1758, and by all subsequent writers. MacGillivray describes them as being very numerous in 1841. Mr Fiddes told Steele Elliot (p. 284) that hundreds passed over the islands in the winter of 1893, great numbers of which perished, and that a few remained over the following summer, but did not breed.

They are extremely abundant at the present day, gathering together on the stone walls and cleits towards dusk ere they seek their sequestered roosts in the cliffs. I saw no evidence, however, of migratory visitors during my sojourns, but the passage of these birds, which usually commences early in October, had not set in ere my visits ended.

* FRINGILLA CŒLEBS, *Chaffinch.*—The first Chaffinch was observed on 5th October 1911, and single birds were seen daily in the crofts until the 9th. On the last-named date small numbers arrived, and some were present on the day of our departure. Seven was the largest number seen in company.

*FRINGILLA MONTIFRINGILLA, *Brambling.*—Several individuals of this rare visitor to the Outer Hebrides arrived during the night of 2nd and 3rd October 1910, and were observed on the stubbles the following day. They evidently passed on at once, for they were not seen afterwards.

In 1911 several appeared on the stubbles on 3rd October, and others followed and were seen daily down to the 12th, the day of our leaving the island.

*CHLORIS CHLORIS, *Greenfinch.*—A few arrived on 10th October 1911, along with other immigrants. A male and female were observed perched on the ridge of one of the houses in the village.

PASSER MONTANUS, *Tree-Sparrow.*—Macaulay (p. 160) records "Sparrows" as having been seen by him during his visit in 1758, and he undoubtedly refers to this species, as the House-Sparrow has not yet found its way to the islands.

Tree-Sparrows are not very numerous, and appear to be confined to the cultivated area and its immediate vicinity, where they breed in the substantial stone boundary walls, and those of the cleits. They were seen in parties during my visits, and were here, as elsewhere in my experience, extremely wary, though entirely immune from any kind of persecution.

ACANTHIS FLAVIROSTRIS, *Twite.*—First came under the notice of MacGillivray (p. 57) in 1840, and said by the Rev. Neil Mackenzie to be a summer visitor to Hirta. I did not find it by any means numerous, and usually only one or two were seen, but on one occasion about a dozen were observed feeding on the stubbles. No increase or diminution in its numbers was noticed, and it was present when we left on 8th October. In

1911, however, it was quite abundant throughout my visit.

* ACANTHIS LINARIA, *Mealy Redpoll.*—Neil Ferguson informed me that shortly after my departure from St Kilda in October 1910, a number of Mealy Redpolls arrived, and remained some time. He frequently saw them on the stone walls, sometimes close to the houses. It is interesting to know that the great invasion of Redpolls, which was such a feature in the ornithology of the autumn of that year, was so far-reaching as to include St Kilda within its scope.

In 1911 one was obtained on 4th October, and another on the 9th. It was the capture of these birds that reminded Ferguson of the visit of this species in the previous year.

* ACANTHIS ROSTRATA (ACANTHIS LINARIA ROSTRATA), *Greater Redpoll.*—We saw several examples of this Greenland bird, and obtained specimens, during our stay on the island in 1911. The first was detected among Twites on 9th September; two were seen on the following day, also in company with Twites; and single birds were captured on the 21st and 29th of the month.

* CARPODACUS ERYTHRINUS, *Scarlet Grosbeak.*—This was one of the unexpected visitors which it was my good fortune to add to our list of novelties. Two males, an adult and a bird of the year, of this North-Eastern European species, were found by the burn to the west of the village on 8th September 1910. On the 19th of the same month an adult female was obtained in some rough herbage amid the crofts.

LOXIA CURVIROSTRA, *Crossbill.*—This bird was added to the St Kildan avifauna by Mr O. G. Pike, who found one alive in the possession of a boy on 7th July 1910 (*Ann. Scot. Nat. Hist.*, 1910, p. 246).

On 17th September we saw a Crossbill along with a small party of Twites in the crofts. On the 20th an adult female, probably the same bird, was observed feeding on ripe thistle-heads just in front of our quarters, and, on being disturbed, flew off with a whole thistle-head in its bill and alighted on the roof. This bird proved to be in excellent condition, and was doubtless one of the band of immigrants which invaded the Scottish isles in the summer. Both the St Kilda birds were examined by me, and were found to belong to the typical Continental form—*L. curvirostra.*

EMBERIZA MILIARIA, *Corn-Bunting.*—The Corn-Bunting was described by MacGillivray (p. 571) as common in July 1840, and as "well-known" by Wilson (ii. p. 72) in 1841; but does not appear to have come under the notice of other observers.

EMBERIZA CITRINELLA, *Yellow Bunting.*—Dixon (p. 83) tells us that he saw one on 18th June 1884.

* EMBERIZA HORTULANA, *Ortolan Bunting.*—Several of these birds, both adults and young, came under our notice, and specimens were obtained. In 1910 they were present on the day of our arrival, 1st September, when two or three were seen in the crofts. Others appeared later, and were observed on the 3rd, 12th, and 16th of the month. In 1911 one only came under notice, namely, on 2nd September. As at Fair Isle, they showed a strong partiality for the standing corn, and were very wary.

* EMBERIZA PUSILLA, *Little Bunting.*—On 15th September 1911, George Stout visited the island of Boreray in company with a number of St Kildans, the object of the voyage being to capture sheep.

Here he saw two Little Buntings on the grass within a few feet of him, but durst not shoot, as the natives and their dogs were chasing sheep all around him. Stout is quite familiar with this bird at Fair Isle, where he has seen as many as a dozen examples in a single season.

* PLECTROPHENAX NIVALIS, *Snow-Bunting.*—The first bird of the season, in 1910, to come under our notice was observed near the summit of Mullach Sgail on 28th September. Though this species has not hitherto been recorded for St Kilda, Mr Neil Ferguson told me that it is seen regularly in numbers on the high ground in the late autumn.

Dr Wiglesworth has informed me that he received a specimen which had been obtained on 28th April 1904. He was told that the bird occurred every spring.

In 1911 the first Snow-Bunting appeared on 20th September. Several were seen on the following day, after which these birds were observed almost daily, sometimes in small flocks, down to the time of our departure.

* CALCARIUS LAPPONICUS, *Lapland Bunting.*—I was particularly pleased to meet with this old friend—one which I have now fallen in with on various Scottish islands for seven Septembers in succession. It was not observed until 20th September 1910, but it may have been present for some little time, since it was found among the crags on the east face of Connacher, at about 800 feet above sea-level. Others were seen on the cliffs facing the Dun Passage on the 21st and 23rd; and on the 24th an adult female was found in the crofts. This bird has only been recorded for the Outer Hebrides,

from the Flannan Isles, where Mr Laidlaw and I found it common in September 1904.

In 1911 this was one of the first birds that came under our notice, and was probably present previous to our arrival on 2nd September. It was seen throughout our visit, being most frequent on rough grass on the high ground, and only occasionally in the crofts. Twelve was the largest number seen in company. Three were seen by George Stout on Boreray on 15th September.

ALAUDA ARVENSIS, *Skylark.*—"Larks" are mentioned by Macaulay (p. 160) as seen by him in 1758, but these may have been Rock-Pipits, a species not included in his observations. MacGillivray (p. 57) describes them as common in July 1840, but no other observer makes any allusion to this bird.

This species visits St Kilda on passage, but it had not put in an appearance up to the date of my departure, 8th October, in 1910.

The first Skylark to come under our notice arrived on 9th October 1911, and was followed by several others on the 10th. This small party frequented the stubbles, and were still present when we left the island on the 12th.

MOTACILLA ALBA, *White Wagtail.*—This Wagtail was not detected as a St Kildan bird until June 1902, when Wiglesworth (p. 40) saw a pair from the 18th to the 25th of that month. Later he received a specimen which had been captured in April 1904. The Rev. Mr Waterston saw one in the Glen on 24th June 1905.

This I found to be one of the commonest birds of passage. It was present on my arrival on 1st September 1910, and was observed daily down to the

23rd of that month. On the 2nd as many as thirty were seen in a party, and there were present, in addition, many scattered birds. A considerable number of fresh arrivals appeared on the 12th, but only a few were seen after the 14th. In 1911 it occurred at intervals down to 3rd October, being numerous from 2nd to 11th September, and again on the 19th and 21st. Her Grace the Duchess of Bedford tells me she saw a number on the island on 23rd August 1910.

* MOTACILLA BOARULA, *Grey Wagtail.*—Three examples were observed, all of which appeared singly on the following dates: an adult, a male, on 1st October; a female on the 3rd; and, lastly, one on the 6th. All were extremely wild, and baffled for a time our efforts to get within reasonable distance of them.

ANTHUS PRATENSIS, *Meadow-Pipit.*—Is described as being common by MacGillivray (p. 57) in July 1840. It is also alluded to by Wilson as occurring on the 2nd and 3rd of August 1841, and by Milner in June 1847. Dixon (p. 83) tells us that it occurs sparingly, breeds, and is said to be resident. On the other hand, Wiglesworth (p. 41), a most accurate observer, saw only a single bird on Mullach Mhor in June 1902, and remarks that it probably breeds, though its nest has apparently never been taken; while other ornithologists who have visited St Kilda make no mention of it.

The above statements are of importance, for we found the Meadow-Pipit extremely abundant throughout my visits—indeed, it was much the commonest bird of passage observed. The Duchess of Bedford observed it on 25th August 1910, and it was present in numbers down to the day of my departure on 8th October 1910, and on the 12th in 1911.

*Anthus trivialis, *Tree-Pipit.*—Quite a number were present on 1st September 1910, and several made their appearance on six subsequent dates, the last on 6th October. It was not observed in the autumn of 1911.

During their visits they chiefly resorted to those portions of the crofts which were under grass.

Anthus obscurus, *Rock-Pipit.*—Macaulay's "Larks," as already mentioned, were probably Rock-Pipits. MacGillivray (p. 57), in 1840, described it as common under the name of *Anthus aquaticus*, and it appears in the lists of all the subsequent visitors interested in birds.

It was quite common during my visits, especially in close proximity to the houses, and very tame.

*Anthus pennsylvanicus, *American Pipit.*—A young male was captured on a small burn close to the village on 30th September 1910. It is doubtful if we should have detected it, in its autumn dress, among the numerous Meadow-Pipits, if its unfamiliar note had not attracted our attention. In this stage of plumage the upper surface much resembles that of a Rock-Pipit, but the under surface is cinnamon brown, the pale markings on the outer pairs of tail feathers are pure white, and the tarsus and toes are black. This bird is new to the British avifauna, and must be regarded, along with the Marsh-Warbler, as the most remarkable and unexpected of our captures in 1910.

*Anthus cervinus, *Red-throated Pipit.*—On 21st September 1910, one flew over my head uttering its unmistakable note, vividly recalling the occasion on which I first made the bird's acquaintance at Fair Isle under precisely similar circumstances on 3rd October 1908.

One was seen at very close quarters on 8th October 1911 (Sunday). It was perched on a wall immediately behind the village, and our attention was drawn to it by its characteristic note.

This bird has only previously been recorded for Scotland from Fair Isle in the autumn of 1908.

* SYLVIA NISORIA, *Barred Warbler.*—This was the first bird of interest that came under notice in 1910. A few minutes after we had landed, one flew out of the Manse garden and alighted on the rocks close to the water's edge. Here it was bullied by a Rock-Pipit, and compelled to beat a retreat, and in doing so it flew past us at close quarters. Unfortunately we did not see it again, as we were busily engaged superintending the landing of our stores, etc.

* SYLVIA ATRICAPILLA, *Blackcap.*—An adult male was captured by a clever bird-catching cat in the Manse garden on 10th October 1911, and brought to me for identification.

* SYLVIA SYLVIA, *Common Whitethroat.*—This was one of a number of interesting visitors which appeared at St Kilda on 6th September 1910. A young male was then found in a plot of potatoes, and was the only representative of the species that came under our notice.

* SYLVIA CURRUCA, *Lesser Whitethroat.*—This species was observed on four occasions in 1910—namely, on 13th, 14th, 15th, and 17th September. Single birds only were detected, except on the 14th, when several came under notice in the crofts. Hitherto a few instances only are known of the visits of this species to the Outer Hebrides. In 1911 a solitary bird was seen on the 7th of October.

* SYLVIA BORIN, *Garden-Warbler.* — This species was observed in small numbers on passage from 2nd to 21st September 1910. During this period it came under notice on seven days, and was especially in evidence, like the Willow-Warbler and other migrants, on the 6th. These migrants frequented the undercliff fringing the beach, the crofts, and the detached enclosures in which grass is grown. Here they were usually to be seen clinging to the heads of the cow-parsnips, which flourished exceedingly on the island, and harboured an abundance of insect food.

SYLVIA SUBALPINA, *Subalpine Warbler.*—One was observed on 13th June 1894, and was shot on the following day in the Manse garden by Steele Elliot (p. 95). The sex of the specimen is said not to have been determinable. A gale of wind had been blowing from the south-west on the day previous to its appearance.

* PHYLLOSCOPUS TROCHILUS, *Willow-Warbler.*—This was one of the most frequent visitors that came under our notice, being observed on no fewer than fourteen days between 1st and 26th September 1910. These birds usually appeared in small numbers, but on the 6th they were fairly numerous. They were found chiefly in the crofts and the detached enclosures, where they sought for insects on the cow-parsnips and other weeds which flourished among the grass. In 1911 one was observed as late as 6th October.

PHYLLOSCOPUS COLLYBITA, *Chiffchaff.*—Several were seen in the crofts on 6th September 1910, and another was observed and obtained on the 12th of the same month. A single bird was obtained on the 11th of

October 1911, and proved to be a specimen of the northern and eastern European race, *P. c. abietina.*

* ACROCEPHALUS PALUSTRIS, *Marsh-Warbler.*—The appearance at St Kilda of this species was a great surprise, as it only occurs regularly in Britain as a local summer visitor to a few counties in Southern England. On 6th September 1910, this stranger was disturbed in a patch of oats, flew a short distance, and alighted in a mass of nettles, where it was secured, and proved to be a female Marsh-Warbler. This is the first example known to have occurred in Scotland, and St Kilda is the most northern and western locality in which the bird has been observed in Europe.

TURDUS MUSICUS, *Song-Thrush.* — MacGillivray (p. 57) tells us that during his visit in July 1840, he "often heard the loud clear song of the Thrush (*Turdus musicus*) resounding along the hillsides"; and Milner (p. 2061) mentions it among the birds that came under his notice on 14th June 1847. The Rev. Neil Mackenzie (p. 76), who was minister on the island from 1829-1843, has recorded in his notebook that this species is a regular winter visitor. It had not, however, put in an appearance before my departure in 1910 or 1911.

TURDUS ILIACUS, *Redwing.*—Dixon (p. 80) states, but without affording any authority, that this species is seen on migration in May and September.

The above statement may be correct. In 1910 we saw the first bird of the season on 20th September.

In 1911 the Redwing was first observed on 23rd September, and a few were seen later in the month. Considerable arrivals took place on 3rd October, after which date it was present in numbers, in various parts of the island, down to our departure on the 12th of the month.

* TURDUS PILARIS, *Fieldfare.*—On 4th October 1911, a party of seven were seen, also two single birds. On the 11th one was observed in company with Redwings on rough ground behind the village. It is probably an annual visitor, but has not hitherto been recorded for St Kilda.

TURDUS MERULA, *Blackbird.*—The Rev. Neil Mackenzie informed Wilson (ii., p. 72), in 1841, that Blackbirds were occasional visitants, and in his notes described it as a regular winter visitor (Mackenzie, p. 76).

A young male of the year was observed by me in the crofts on 6th October.

* ERITHACUS RUBECULA, *Redbreast.*—Neil Ferguson informed me that this bird is an annual autumn visitor, and that he sees it about the houses during the winter. He described the bird well.

* RUTICILLA PHŒNICURUS, *Redstart.*—This species appeared on two occasions in 1910: on 12th September a bird of the year was observed; and on the 26th two were seen on the rocks near the landing-place, and another in the crofts. There are a few records only of the visits of this species to the Outer Hebrides, and none previously for St Kilda.

SAXICOLA ŒNANTHE, *Wheatear.*—Martin (p. 46), in the lists of birds observed by him in 1697, mentions this species under the name of "Stonechacker." MacGillivray (p. 57), the next observer to allude to the Wheatear, states that it was breeding plentifully among the stones in 1840. Steele Elliot (p. 285), in 1894, estimates the pairs breeding at six only.

We saw several examples between 1st and 5th September 1910, but did not detect this typical form later, though we were continually on the look-out for it. In 1911, however, one was observed as late as the 21st.

* SAXICOLA LEUCORRHOA (SAXICOLA ŒNANTHE LEUCORRHOA), *Greater Wheatear.*—This large north-western race of the Wheatear was constantly observed on passage, frequently in abundance, from 2nd September down to 12th October, and possibly occurred later. It was especially numerous on 12th to 13th September 1910, and the latest date for its appearance in numbers was 4th October 1911. This was one of the commonest birds of passage, and was observed on no fewer than twenty-eight days during my sojourn. The wings of males obtained varied from 100-108 mm., and those of females from 100-107 mm.

In 1911 it was present throughout our visit in varying numbers. The latest date recorded for its arrival in some abundance was 4th October.

* SAXICOLA HISPANICA, *Black-throated Wheatear.*—On 21st September 1911, we were fortunate enough to discover an example of this rare visitor to the British Isles on high ground over which large boulders were scattered. Our attention was drawn to the bird by its small size and pale colour. It proved to be a female, which had recently assumed its winter dress. This elegant summer visitor to the western portion of the Mediterranean basin had only once previously been obtained in Scotland, namely, by myself at Fair Isle in September 1907 (see p. 145).

* PRATINCOLA RUBETRA, *Whinchat.*—A female seen in the crofts on 8th September 1910 was the only bird of this species to come under notice.

PRATINCOLA RUBICOLA, *Stonechat.*—All that we know about the Stonechat as a St Kilda bird is contained in the brief statement made by the late Howard Saunders in his *Manual of British Birds*, p. 29, that he "observed it on St Kilda in August 1886."

TROGLODYTES HIRTENSIS (TROGLODYTES TROGLODYTES HIRTENSIS), *Wren.*—This bird came under the notice of Martin (p. 46) in 1697; and Macaulay (p. 160) also observed it sixty-one years later. The latter author remarks, "How these little birds—I mean the Wrens particularly—could have flown thither, or whether they went accidentally in boats, I leave to be determined."

The marked peculiarities in the plumage of the St Kilda Wren, and its large size as compared with other British examples, remained unnoticed until 1884, when Seebohm (*Zoologist*, 1884, p. 333) described it as a new species, under the name of *Troglodytes hirtensis*, from specimens obtained by Dixon in that year. It breeds on Hirta, Soay, Dun, Boreray, and Stack an Armin.

We found the Wren in all parts of Hirta, among the boulders that fringe the head of the bay, in the walls and cleits, among the crofts, on the screes and rocks on the hillsides, and in the faces of the great cliffs. Three nests were shown us. One of these was placed in a hole worked in a mass of dead thrift on the face of a cliff; the other two were placed between the stones forming the inner walls of cleits, and were in excellent condition. All these nests were composed of the blades and stems of grasses, small tufts of grass, a little moss and dead bracken, and were lined either entirely with white feathers of a gull, or with a mixture of moss and white feathers.

HIRUNDO RUSTICA, *Swallow.*—The Swallow is described by the Rev. Neil Mackenzie (p. 76), who resided in the island from 1829 to 1843, as an occasional visitor. It seems, from the scanty data, to be irregular in its appearance, but in the years 1883 and 1885 Swallows were seen, according to Dixon (pp. 83 and 360), in late spring in some numbers. Dr Wiglesworth informs me that one was found alive in the church on 29th May 1907. There are several other records, all relating to a few birds, seen during the first half of the month of June.

CHELIDON URBICA, *House-Martin.*—The visits of single specimens of the Martin have been noted on four occasions—namely, on 9th June 1887, as recorded in *The Ibis* (1887, p. 470) by the Rev H. A. Macpherson, on the authority of Mr James Murray; one obtained by Mr Mackenzie, the factor, on a date unmentioned (Steele Elliot, p. 284); one on 17th June 1905; and another a week later, seen by the Rev. J. Waterston (*Ann. Scot. Nat. Hist.*, 1905, p. 202).

* IYNX TORQUILLA, *Wryneck.*—One was found resting on the stone wall which forms the western boundary of the crofted area, on 6th September 1910, the most productive day for migrants during our sojourn on the island in that year.

CUCULUS CANORUS, *Cuckoo.*—The visits of the Cuckoo to St Kilda have always been regarded by the inhabitants as the precursor of important events. Martin (pp. 46-47) tells us that it is "very rarely seen here, and that upon extraordinary occasions, such as the death of the proprietor Macleod, the steward's death, or the arrival of some notable stranger." The same beliefs still survive, and were reported in the public press in connection with the bird's appearance in the spring of 1910. The Rev. Neil Mackenzie (p. 76) records it as being a rare visitor in his time (1829-1843), and repeats Martin's statements regarding the superstitions the natives associate with its visits. The bird does indeed appear to be a rare visitor, for there is no other information regarding its appearances on record.

CYPSELUS APUS, *Swift.*—Single examples have been noticed on four occasions: one by Mr John Mackenzie in May 1886 (H. A. Macpherson, *The Ibis,* 1887, p. 470); another by Mr Mackenzie, but no date given (Steele Elliot, p. 284); one by Mr Wheat in 1899; and one on 8th June 1902 (Wiglesworth, p. 40.)

CORACIAS GARRULUS, *Roller.*—During his visit to St Kilda in August 1841, Wilson (ii., p. 73) was informed by the Rev. Neil Mackenzie of the appearance, some time previously, of a rare bird which remained on the island for several weeks, and which from the description Wilson diagnosed as being undoubtedly a Roller.

HALIAËTUS ALBICILLA, *Sea-Eagle.*—It is greatly to be regretted that St Kilda, with its ideal haunts for this fine species, no longer knows the Sea Eagle as the King of its birds. Martin (p. 46) alludes to the presence of "Eagles" in 1697, and in his *Western Isles of Scotland,* published in 1703 (p. 299), tells us that there is "a couple of large eagles who have their nest on the north end of the Isle; the inhabitants told me that they commonly make their purchases on the adjacent isles and continent, and never take so much as a lamb or a hen from the place of their abode, where they propagate their kind." Macaulay (pp. 160-161), who visited St Kilda in 1758, remarks: "At Hirta are a few Eagles, which, though very pernicious elsewhere, are perfectly harmless here, the reason I conceive must be that their necessities are more than sufficiently supplied by the inexhaustible stores of eggs that must every moment fall in their way. This must be the case in summer. How they procure their food in winter is a question which one will find some greater difficulty in resolving, unless we take it for granted that they make frequent excursions into the neighbouring isles." Unfortunately we do not know the date of the banishment of this bird from the isles so eminently suited for its home, but Wilson (ii., p. 72), who visited St Kilda in 1841, tells us that "there are now no Eagles either on the main island or its dependencies." Milner, however, mentions it as one of the birds seen by his party on 15th June 1847, perhaps on the word of the notorious David Graham, of York, who formed one of the company. The Rev. Neil Mackenzie (p. 76) describes it as an occasional visitor, which would seem to indicate that it had ceased to be a native before his induction to the island in 1829. Steele Elliot (p. 282) tells us that the site of the eyrie was on the Connacher cliffs, whose height is 1260 feet.

*FALCO CANDICANS, *Greenland Falcon.*—Finlay McQueen informed me that in the early spring of 1910 he surprised, by coming upon it suddenly, a Falcon much larger than a Peregrine, and of a pure white colour with dark markings. There can be no doubt that the bird seen was a Greenland Falcon—a species which was unusually numerous in various parts of Scotland in the early months of the year named.

FALCO PEREGRINUS, *Peregrine Falcon.*—Martin (p. 46) writes: "Hawks extraordinary good;" and in his *Western Isles of Scotland* says, that "this Isle produces the finest hawks in the Western Isles, for they go many leagues for their prey, there being no land-fowl in St

Kilda proper for them to eat, except pigeons and plovers." The Rev. Neil Mackenzie (p. 81) remarks that they remain all the year round. The Goshawk, *Falco palumbarius* of Atkinson (p. 224), is undoubtedly this species.

At the present time, I was informed that two, or at the most three pairs nest in the St Kilda group, and that the sites of the nests are on Boreray, Dun, and perhaps Soay. We frequently saw these birds during our rambles.

FALCO ÆSALON, *Merlin.*—Steele Elliot (p. 282) describes it as occurring frequently, and relates that one flew in at the Manse window and was secured. There are no other records, however.

We saw one or more of these birds almost daily throughout our visits. They proved a great source of annoyance to us, since they frequently swept over the crofted area, our main observing ground, in search of small birds, among which they naturally created great alarm and unrest. This species is probably a regular visitor to the islands during the periods of the great migratory movements in the spring and autumn.

FALCO TINNUNCULUS, *Kestrel.*—MacGillivray (p. 56) in 1840 states that it breeds in the precipices in small numbers; and Wilson (ii., p. 72) includes it as a member of the St Kilda avifauna in 1841. Dixon (p. 80) says that it is an occasional visitor, but that he could find no evidence of its ever breeding.

Single individuals, probably birds on passage, came under my notice on 8th and 19th September.

[ACCIPITER NISUS, *Sparrow-Hawk.*—Dixon (p. 359) tells us that one was seen by Mr Mackenzie on

7th June 1885, and he gravely suggests that it may breed!]

PHALACROCORAX CARBO, *Cormorant.* — Atkinson (p. 224) includes this species as among those seen by him in May 1831; and MacGillivray (p. 64), who visited St Kilda in July 1840, states that both Cormorant and Shag are found. Milner (p. 2062) tells us that the eggs of both the Cormorant and the Shag were obtained by his party on 15th June 1847. Subsequent visitors have, however, searched for it in vain as a breeding bird.

We saw a number throughout our visits, and the bird is a regular winter visitor, as it is to many other insular stations in which it does not breed.

PHALACROCORAX GRACULUS, *Shag.*—First noted for St Kilda by MacGillivray (p. 64) in 1840, and described by him as being in great numbers.

The same statement holds good at the present time; it is to be seen everywhere, except inland.

SULA BASSANA, *Gannet.*—The British metropolis of the Gannet is on the Isle of Boreray, and the adjoining stacks of Lii and An Armin. Here they have come under the notice of all who have written on the avifauna of the islands since Martin made his memorable visit in 1697. The birds were still on the ledges or summits of the stations named when we surveyed them from the deck of the *Wamba* on 8th and 12th October. A number of young of the year floated, in helpless fashion, on the adjoining seas. The natives of St Kilda no longer capture the young Gannets in the autumn for consumption during the winter, and consequently the birds must have increased considerably during recent years. About 600 adults are, however, captured in spring, for food.

I was much surprised, on leaving St Kilda on 12th October 1911, to witness this species participating in the scramble which took place for entrails and other fish refuse which the cook of the *Mercury* cast overboard when preparing dinner. Yet a number of adult Gannets contested with the numerous Gulls, Pomatorhine Skuas, Shearwaters, and Fulmars for a share of the offal, which, when successful, they devoured with the greatest gusto. I was not previously aware that this species was to be regarded as one of the scavengers of the ocean.

ARDEA CINEREA, *Heron.*—Macaulay (p. 160), writing in 1764, remarks: "Should it be asserted here that the St Kildans have, by dint of stalking, caught Herons, that is the most watchful fowls in the world, I am afraid the story would hardly be credited, though the fact seems to be well attested." The natives of to-day, however, regard this as a yarn, and do not claim such mastery in the use of their "bird-rods." Wilson (ii., p. 81), on the authority of the Rev. Neil Mackenzie, states that the Heron visits the island in the winter. The Rev. J. Waterston (*Ann. Scot. Nat. Hist.*, 1905, p. 202) observed one in June 1905. Mackenzie (p. 76) remarks that the Herons visiting the islands die of starvation, which is likely, since there are no suitable feeding grounds. We observed and heard one or more on many occasions between the 4th September and 2nd October 1910. They came under notice chiefly at night, and made their presence known by their harsh unmusical notes. These birds probably spent the daytime on some of the adjacent islands of the group.

This species was not seen in 1911 until 1st October, when a single bird appeared. Several arrived with the fine weather which set in on the 3rd, and one was observed on the 5th.

CYGNUS SP., *Swans.*—According to the Rev. Neil Mackenzie (p. 76), a few Swans (probably Whoopers) visit the island in October and November in stormy weather, but do not remain.

ANSER ALBIFRONS, *White-fronted Goose.*—An adult male of this species was shot between 6th and 22nd June 1895, and is now in the collection of the Royal Scottish Museum, to which it was presented by the Rev. H. A. Macpherson (*Ann. Scot. Nat. Hist.*, 1895, pp. 252-53). The Rev. Neil Mackenzie mentions that a few "Wild Geese" visit St Kilda during stormy weather in October and November, but seldom remain longer than a few days.

ANSER SP., *Grey Geese.*—Neil Ferguson observed a party of about forty "Grey Geese" flying over the island at 7 A.M. on 22nd September 1911. They were very noisy, and passed in the direction of the island of Soay.

BRANTA BERNICLA, *Brent Goose.*—Heathcote (*St Kilda*, p. 193) mentions that he saw one on Cambargu, the extremity of the north-west peninsula of Hirta. No date, however, is mentioned.

A single bird was observed flying over the bay on 21st September 1911.

ANAS BOSCAS, *Mallard.*—The skin of one shot in September 1902 was sent to Wiglesworth (p. 40). Mackenzie (p. 76) tells us that some are observed in November, but seldom remain longer than a few days.

On 30th September 1911, a female was observed in the bay, and an adult male was found on a small pool on the high ground on 8th October.

* Dafila acuta, *Pintail.*—An immature male was found on 14th September in a small piece of marshy ground formed by the waters of the burn ere it enters the sea at the East Bay. Another was seen in the same place on 7th October 1911.

* Mareca penelope, *Wigeon.*—An adult male was found on a pool near the summit of Mullach Sgail on 24th September 1911; and a female was observed on the Glen burn on 10th October.

* Nettion crecca, *Teal.*—On 13th September a male was on the East Bay, and was afterwards seen feeding in a marshy corner of the crofts. On the 21st, another, or perhaps the same bird, rose from the burn which runs past the village on the west. This species is new to the St Kildan avifauna.

In 1911 single birds were seen on 13th September and 2nd October, and two were seen in the bay on 30th September.

* Harelda glacialis, *Long-tailed Duck.*—Three, an adult male and two females or immature birds, arrived on 7th October 1911 and spent a little time in the bay. It was observed that they did not seem to appreciate the company of the numerous Cormorants and Shags, and they were restless in consequence and did not remain long.

Somateria mollissima, *Eider Duck.*—The Eider Duck would appear not to have been a breeding species in the Rev. Neil Mackenzie's day (1829-1843), for he describes it (p. 76) as being occasionally seen in late autumn. Mr Mackenzie, in 1841, made the same statement to Wilson (ii., p. 80), who, on his authority, relates that it now and then visits the shores in October. Milner (p. 2061), however, includes it among the species observed in June 1847, and Elwes (*Ibis*, 1869, p. 36) found it breeding on Dun in 1868.

I was informed by Neil Gillies that it breeds commonly on the flat tops of Mullach Sgail (705 feet), and Mullach Mhor (1153 feet). During our visit adult males, and females with half-grown young, were seen in the village bay.

SOMATERIA SPECTABILIS, *King Eider*.—Dixon (p. 58) states that two pairs were seen daily by him in the bay, along with Common Eiders, during his visit in June 1884; and he had not the slightest doubt that they were breeding on the island of Dun.

MERGUS SERRATOR, *Red-breasted Merganser*.—Dixon (p. 87) mentions that one was shot in 1883, and that Mr Mackenzie, the factor, shot one during June 1884, in the East Bay. He saw a stuffed specimen. Steele Elliot (p. 285) observed one in the bay on 11th June 1894.

We saw single examples in the East or Village Bay on several occasions after 5th September 1910, and down to our departure on 12th October 1911.

COLUMBA LIVIA, *Rock-Dove*.—Martin mentions in his *Western Isles of Scotland* (1703, p. 296) that the Hawks on St Kilda prey upon Pigeons and Plovers, no doubt basing his information on the knowledge gained during his visit to the island in June 1697. Macaulay (p. 160), who was at St Kilda in 1758, includes "Pigeons" among the birds to be found there. Milner (p. 2061) mentions the Rock-Dove as amongst the species seen by him on 15th June 1847. Dixon (p. 84), writing in 1884, says there can be little doubt that it breeds, but he never saw more than one pair. Steele Elliot (p. 284), in 1894, tells us that he was unable to find any trace of this bird. No one else among the ornithologists who have visited the island makes any allusion to this species. I made enquiries among the natives concerning this bird, and found that none of them had ever seen it. Donald Ferguson, the oldest man in St Kilda, aged 77, informed me that his father had told him that Doves used to breed in a cave near the Point of Coll, the north-east promontory of Hirta. The small amount of land under cultivation, and the limited area of low land, probably render the island little suited for the requirements of this species. The area under the spade is no doubt less now than in times past.

* TURTUR TURTUR, *Turtle-Dove.*—A young female appeared on 13th September 1910. When first seen it was being hotly pursued by a number of Meadow-Pipits, which had evidently mistaken the bird for some species of hawk. The Turtle-Dove has seldom been reported from the Outer Hebrides, and had not previously been recorded for St Kilda.

Dr Wiglesworth writes me that he received an immature bird which had been obtained in September 1902, and that a second was seen at the same time.

In 1911 one was seen on the side of Oisaval on 7th September, and a bird of the year appeared in the crofts, pursued by Pipits, on the 19th.

LAGOPUS MUTUS, *Ptarmigan.* — The Rev. Neil Mackenzie informed Wilson (ii., p. 37), in 1841, that on one winter day he had seen a single Ptarmigan on the hillside, the wind previous to its appearance having been from the east.

CREX CREX, *Corn-Crake.*— This species, under the name of "Craker," has a place among the birds observed by Martin (p. 46) in 1697. MacGillivray (p. 58), who was at St Kilda in the summer of 1840, states that a few occurred among the corn, and that their cry might be heard all night long. It was also heard by Wilson (ii., p. 73) in the first week of August in 1841. These are the only visitors who make any mention of this species, which would seem to have been a summer visitor to the main island in the past. It may have ceased to seek the island when the area under corn was insufficient to offer it inducements to nest, as the grass is not rank enough to afford the bird suitable cover.

* RALLUS AQUATICUS, *Water-Rail.*—A specimen captured by a cat on 3rd November 1903, and sent to Dr Wiglesworth, is the only known instance of the occurrence of this species at St Kilda.

* FULICA ATRA, *Coot.*—This was a most unexpected visitor to meet with at such a far-western isle. It was found on 24th September 1910 at the foot of the low cliff which encircles the head of the East Bay. It appeared to be fatigued, and took wing unwillingly when pressed, to alight again a few yards off. Finally, on being further disturbed, it flew some distance out into the bay, alighted in a calm sea, and was not seen again.

Since my return, Dr Wiglesworth has informed me that he received an immature specimen which allowed itself to be picked up, and that another was seen on the water at the same time—*i.e.*, 12th November 1902.

* ÆGIALITIS HIATICOLA, *Ringed Plover.*—This species is not a native bird; indeed it does not appear to have been observed previously in the islands. During our

visits a few were present throughout September and down to 11th October, frequenting the hillsides or the beach and its vicinity. Some of these seemed to be family parties, and remained for a few days. On the 11th of September 1910, a flock containing about a score arrived, and tarried until the 15th, when they passed on.

CHARADRIUS PLUVIALIS, *Golden Plover.*—Martin (p. 46), in 1697, and Macaulay (p. 160), in 1758, mention "Plovers," the former stating in his *Western Isles of Scotland* (p. 296) that Plovers and Pigeons form the food of the St Kilda Hawks. MacGillivray (p. 58), in 1840, was informed by the Rev. Neil Mackenzie that Golden Plovers occasionally occurred. Heathcote (p. 193) tells us that a party used to frequent the top of Mullach Mhor. No date is mentioned for this latter statement, but it was probably in the late summer.

On 2nd September 1910, the first day on which we visited the high ground, we saw five of these birds, and we observed them in similar haunts throughout our visits, often in considerable numbers.

In 1911 it was numerous on the high ground from 5th September to our departure. There was a big arrival on 30th September.

* SQUATAROLA HELVETICA, *Grey Plover.*—George Stout saw one in company with Golden Plovers on 28th September 1910, on the extensive flat which forms the top of Mullach Sgail. When disturbed, the bird flew off alone. Stout is quite certain as to the identity of his bird. The Grey Plover appears to be a somewhat rare visitor to the Outer Hebrides.

VANELLUS VANELLUS, *Lapwing.*—The Rev. Neil Mackenzie informed Wilson (ii., p. 81), in 1841, that

Lapwings visited the island in winter; and in 1894 the Rev. Mr Fiddes assured Steele Elliot (p. 84) that the bird occurred every spring on migration. The Rev. J. Waterston saw one on Ruaval on 19th June 1905 (*Ann. Scot. Nat. Hist.*, 1905, p. 202).

* STREPSILAS INTERPRES, *Turnstone*.—A few birds of this species may possibly be winter visitors, but we did not observe any Turnstones after 30th September, though down to that date a few had been present throughout our visits. The rocks fringing the East Bay, including those of Dun, were their chief resorts. They were not numerous, and never more than five were seen together.

HÆMATOPUS OSTRALEGUS, *Oyster-catcher*. — Martin (p. 64) mentions this species as the "Tirma or Sea-pie" —and Macaulay (p. 160) as the "Sea-Magpie." Since the days of these earliest writers, the bird has come under the notice of most visitors to St Kilda. The Rev. Neil Mackenzie (p. 153) says that it arrives in February, and leaves when the young ones are fully fledged, and that a few remain all winter. Wiglesworth (p. 52) says that it is not numerous, but that several pairs breed round the coasts.

We only saw single birds on 3rd and 8th September 1910, and a single bird was present on our arrival, and remained for several days. Two were seen on 8th September, and one on 1st October.

Neil Ferguson informed me that considerable numbers breed, but leave the island during August with their young.

SCOLOPAX RUSTICULA, *Woodcock*.—The Rev. Neil Mackenzie (p. 76) says that a few generally come in November, but not regularly. Neil Ferguson told me that in some seasons it is found in numbers on the hills in late autumn.

On 5th and 6th October 1911, several were observed on the rough, rocky faces of the hills; and on the 9th one which had been disturbed on the hillside flew down into the crofts. No doubt many more were present than came under our notice, but the nature of the ground the birds frequented, and its great extent, made them very difficult to find.

* GALLINAGO MAJOR, *Great Snipe.*—This was another of the interesting visitors which came under observation on 6th September 1910. In beating the crofts on that day, one was flushed from some rough grass, and alighted among a patch of standing oats. The occurrence of this species at St Kilda is of special interest, as the bird does not appear to have been hitherto observed in any of the Outer Hebridean islands.

GALLINAGO GALLINAGO, *Common Snipe.*—Although the Rev. Neil Mackenzie (p. 76), who was minister on St Kilda from 1829 to 1843, avers that a few were resident, yet it was not until the year 1900, as we are informed by Wiglesworth (p. 41), that the first eggs were found. Since then other nests have been obtained, and it probably breeds annually. Waterston (p. 202), heard one drumming on Mullach Mhor on 14th June 1905.

We found the Snipe singly in the crofts (sometimes among corn), and on the higher ground, throughout our stay in 1910. On 10th September a party of natives who visited Soay, found this bird swarming among the long grass which in places clothes that island. One of these birds, which the captor had taken by putting his foot upon it, was brought to me for determination.

In 1911 it occurred at intervals between 16th September and 11th October, with considerable arrivals on 4th and 6th October.

TRINGA ALPINA, *Dunlin.*—MacGillivray (p. 58), in July 1840, saw "several pairs on the hillside, where they doubtless had young." On 14th June 1902, Wiglesworth (p. 41) saw four in full breeding plumage in Village Bay, and remarks that it "possibly breeds."

Only a single bird came under my notice in 1910. This was seen in company with Sanderlings, on the strip of sandy beach below the village on 2nd September. In 1911 a single bird was present from 5th to 9th September, and two were seen on the 10th. This common and ubiquitous wader seems to be a *rara avis* at St Kilda.

* TRINGA BAIRDI, *Baird's Sandpiper.*—On 28th September 1911, a small wader was observed in a pool on the rocks fringing the bay. It was shot on the suspicion that it was something uncommon, and proved to be an adult female of this North American bird in full winter plumage. This species had only on two previous occasions been known to visit the British Isles, but had not before been detected in Scotland. Baird's Sandpiper breeds on the Arctic coast of America from Bering Strait to the northern shores of Hudson Bay, and winters in Chili, Argentina, and Patagonia.

* TRINGA CANUTUS, *Knot.*—The wing and other remains of a bird of the year, which had been captured by a cat on 10th September 1910, were brought to me for identification, and were the first evidence obtained of the occurrence of this species at St Kilda.

In 1911 single birds were seen on the rocks fringing

the bay on the 2nd and 8th of September, and two on the 7th of the month.

* CALIDRIS ARENARIA, *Sanderling*.—In 1910 a party of three birds in immature dress were present on the sandy beach, at the head of the East Bay, on the day of our arrival, 1st September, and remained until the 6th. None were seen later. This species is probably a regular bird of passage in both spring and autumn.

* MACHETES PUGNAX, *Ruff*.—A male was seen on the grassy slopes above the cliff on the south side of the bay from 3rd to 6th September 1910. Another, or possibly the identical bird, was observed on the same ground on the 13th.

TOTANUS HYPOLEUCUS, *Common Sandpiper*.—This species may be a more or less regular visitor during the spring and autumn migrations to and from its northern summer haunts. Milner (p. 2061) saw one at St Kilda on 14th June 1847; and Her Grace the Duchess of Bedford informs me that she observed a pair there on 23rd August 1910.

In 1911 we saw one on the rocks bordering the bay on the 2nd of September, and another on the 6th.

* TOTANUS CALIDRIS, *Redshank*.—I had expected to come across the Redshank on my first visit, but the bird seems to be an uncommon species at St Kilda. The only known instance of its appearance was on 10th September 1911, when two of these birds were seen, and heard uttering their unmistakable call-notes, as they flew across the bay towards the island of Dun.

NUMENIUS ARQUATA, *Curlew*.—Macaulay (p. 160) enumerates "Curlews" amongst the birds of St Kilda in 1758. The Rev. Neil Mackenzie informed Wilson (ii., p. 81), in 1841, that this species was a winter visitor to the island. Dixon (p. 85) tells us that he saw two pairs in June 1884, and thinks that they must have had nests on the island of Dun or in Glen Mor, but could not find them. This species does not, however, appear to have come under the notice of other ornithologists visiting the islands.

Curlews on autumn passage were seen in varying numbers throughout my visits, being observed daily from 2nd September to 12th October. Sometimes only one or two happened to come under notice, but on 5th September 1910, a dozen were observed in company, and this was the largest number seen. On 17th September several were feeding on the sides of Oiseval, one of which several times uttered the pretty summer note so characteristic of the species—much the latest date on which I have ever heard it.

NUMENIUS PHÆOPUS, *Whimbrel.*—Dixon (p. 85) saw a pair in June 1884; and Wiglesworth (p. 41) describes it as not uncommon round the coasts, where he frequently saw it in June 1902. The latter writer adds that he failed to discover the least evidence of its breeding. Steele Elliot (p. 284) saw six of these birds in the East Bay on his arrival, and these were still there at the time of his departure on 21st June 1894, but he too saw no trace of any nesting-haunt. Waterston (p. 202) says that five or six pairs were resident in the summer of 1905, and that one native collector told him that he had taken and sold the eggs.

I was somewhat surprised not to have observed this species in 1910, but it may have kept to the extensive tracts of high ground and thus have escaped notice.

In 1911, however, we saw single birds on 3rd, 7th, and 8th September. This is one of the species that we should expect to be a regular visitor on its spring and autumn movements between its seasonal haunts, which include Iceland in summer.

*STERNA MACRURA, *Arctic Tern*. — The Duchess of Bedford informs me that she saw an Arctic Tern fishing in the East Bay near the landing-place on 23rd August 1910. Terns would appear to be extremely rare in the St Kildan seas, for they have not hitherto come under the notice of any of the naturalists who have visited the group. One would have expected some of these birds to appear annually during the spring and autumn movements to and from their northern breeding-haunts.

LARUS RIDIBUNDUS, *Black-headed Gull*.—The Rev. H. Macpherson (*Ibis*, 1887, p. 470) was informed by Mr George Murray that one arrived on St Kilda on 13th April 1887; and Steele Elliot (p. 286) mentions that a specimen of this Gull was once obtained by Mr J. Mackenzie. I am informed by Dr Wiglesworth that he received a specimen which had been caught on the sand at the head of the East Bay. Twelve others were observed on the water close by.

LARUS CANUS, *Common Gull*.—J. MacGillivray, who was at St Kilda in July 1840, describes this species as the least common species of Gull, and Milner (p. 2062) records that a nest and eggs of the Common Gull were taken at Boreray, St Kilda, on 15th June 1847. Kearton (*With Nature and a Camera*, p. 57) includes this species in the list of birds he "saw and identified in the islands," but as the Kittiwake has no place in this list, the inference is obvious. On the other hand, Dixon

(p. 87), Steele Elliot (p. 286), and Wiglesworth (p. 42) all failed to find it.

I thought it was quite likely that I should see this and other species not natives of the islands, during the autumn, when so many Gulls are on the move and have a much wider range than at any other season; yet, although a very careful look-out was kept, only one, an immature bird, came under notice. This was observed feeding on the land in front of our house on 7th October 1911. It would seem that the southern movements of certain Gulls and Terns do not extend regularly as far west as the St Kilda group.

LARUS ARGENTATUS, *Herring-Gull.*—This is no doubt the middle-sized Gull alluded to by Martin (p. 63) as one of the three species of "Sea-Malls" inhabiting St Kilda in 1697. It is first alluded to as the Herring-Gull by Atkinson in 1839, and is mentioned by all other ornithologists who have treated of the birds of this group of islands.

It is apparently not abundant in the breeding season, according to Dixon (p. 86) and Wiglesworth (p. 53), but it breeds on all the islands.

In the autumns of 1910 and 1911 we found both adults and birds of the year numerous throughout our visits.

LARUS GLAUCUS, *Glaucous Gull.*—This species is, no doubt, a regular winter visitor, arriving in late autumn and departing in early spring. At present, however, we only know of a single instance of its occurrence—namely, in November 1886, when one was seen in the East Bay, and recorded by the late Rev. H. A. Macpherson, *Ibis* for 1887, p. 470, on the authority of Mr George Murray.

Since the above was written, Dr Wiglesworth tells me

that he received an adult specimen which had been obtained on 7th May 1903.

*LARUS LEUCOPTERUS, *Iceland Gull.*—An immature example, captured in March 1903, was forwarded to Dr Wiglesworth. This is the only known instance of the occurrence of the species.

LARUS FUSCUS, *Lesser Black-backed Gull.*—MacGillivray (p. 59) includes this Gull among the species seen by him in July 1840. It is not, however, mentioned by Wilson, an accomplished naturalist, who visited the islands in the spring of the following year. Milner (pp. 2059-2062) tells us that in June 1847, the Greater and Lesser Black-backed Gulls were breeding on Dun in abundance. Dixon (p. 86) says that it breeds in considerable numbers on St Kilda and the adjoining islands and stacks; but Steele Elliot (p. 95), Wiglesworth (p. 41), and others, have been quite unable to find it anywhere in the archipelago during the nesting season.

In 1911, however, I heard that this species bred at St Kilda during the summer. A single nest, with eggs, was found by Neil Ferguson, who told me that this is the only instance known to him of this Gull having bred on any of the islands of the St Kilda group.

We were always on the look-out for this species, but were only rewarded by seeing two birds of the year, in company with two young Herring-Gulls, on the sand at the head of the East Bay on 1st September 1910, and another on 10th September 1911.

LARUS MARINUS, *Great Black-backed Gull.*—This is the largest form of "Sea-Mall" mentioned by Martin (p. 63) as inhabiting St Kilda in 1697, to which he gives the native name of "Tuliac." Macaulay (p. 158) tells us that this bird is hated by every St Kildan, on account of its being a destroyer of young sea-fowls and of eggs, and if caught, "they outvie one another in torturing this imp of hell to death."

It is a resident species, and fairly plentiful. We saw both young and old birds in numbers throughout our visits.

RISSA TRIDACTYLA, *Kittiwake.*—This, by far the most abundant of the St Kildan Gulls, is, no doubt, the third and smallest of the "Sea-Malls" mentioned by Martin in his 1697 list, where it is named "Reddag," and is described as being white and less than a tame duck, and as arriving on 15th April and departing in August. The accuracy of the dates given by Martin is confirmed by other observers.

Although the Kittiwake quits its nesting-haunts on the cliffs during August, yet it by no means leaves the vicinity of the islands, and its peculiar cries were often heard throughout our visits. During September 1911 young birds were numerous in the bay. There were hundreds of these birds close to the cliffs of Boreray when we steamed under them on 8th October 1910.

MEGALESTRIS CATARRHACTES, *Great Skua.*—Dixon (p. 87) remarks that this species appears on St Kilda at irregular intervals, but he quotes neither instances nor authority for the statement.

I am inclined to think that it visits the seas in the immediate vicinity of the islands at regular periods. We saw one close to Boreray, and six others at sea, on 8th October 1910; and one was observed off Boreray on 12th October 1911. It probably also occurs in the spring.

* Stercorarius Pomatorhinus, *Pomatorhine Skua.* —On the 12th and 13th of October 1911, this species was observed in numbers all the way from St Kilda to near Stornoway. It scrambled along with Gannets, Shearwaters, etc., for the odds and ends of fish which the cook of the *Mercury* cast overboard, and plunged into the water, completely immersing itself, to secure a share of the sinking tit-bits.

Stercorarius crepidatus, *Arctic Skua.*—Dixon is the only previous visitor to mention this species. He tells us (p. 87) that he was informed that an example was observed on St Kilda in the summer of 1883.

No doubt it frequently occurs in the vicinity of the islands, during the period of its spring and autumn passages. We saw several in attendance on the Kittiwakes at Boreray on 8th October 1910, and one came under our notice there on 12th October 1911.

Alca torda, *Razorbill.*—This very common breeding species is described by Martin (p. 61) under the local name of "Falk." In the *Zoologist*, 1887, p. 347, Macpherson records that an example with pied upper plumage was captured at St Kilda in June 1887.

The Razorbills had left their haunts on Hirta some time before our arrival, but we saw many at sea, especially off Boreray, when leaving the islands on 8th October 1910.

Alca impennis, *Great Auk.*—St Kilda holds a unique place in the British history of this very remarkable bird. It was from here that the "Garefowl" was made known to us as a native of our islands; and it was here, it is believed, that the last of the British race met its death as a victim of superstition. It is from the St Kilda group, too, that we have obtained more of the scanty knowledge we possess relating to the unfortunate Great Auk than from any other quarter of the British seas.

In the account of Hirta furnished by Sir George Mackenzie, the Lord Clerk Register, to Sir Robert Sibbald, mention is made of the incredible number of birds to be found there, amongst which there is one they call the "Garefowl," which is "bigger than any goose, and hath eggs as big almost as those of the Ostrich." This account unfortunately bears no date, but, thanks to the investigations specially made for me by my friend Mr Henry Johnstone, it is possible to fix it as having been written between the years 1681 and 1685, for Sir George was made Lord Clerk Register in the former year, and was created Viscount Tarbat in the latter. The last of the British race of Garefowls was captured on Stack an Armin, off Boreray, in the month of July, about the year 1840, and done to death as a witch. This, however, is not the place to discuss such an important subject as the history of the Great Auk as a St Kilda bird, but I should like to point out that Mr G. C. Atkinson, who was at St Kilda on the last day of May 1831, gives (p. 224) a list of "the birds which we observed on St Kilda," and among them is the "Great Auk, *Alca impennis*." Unfortunately, Mr Atkinson seems to have been chiefly interested in the food-birds of the inhabitants and their methods of capture. The extreme importance, however, of observing a Great Auk was not then realised.

The Rev. Neil Mackenzie (p. 75), who was minister at St Kilda from 1829 to 1843, informed his son that though "he made all possible enquiry, none of the natives then living had ever seen it, but they had heard of a bird of that kind, which they vaguely described. After consideration of all that he could ascertain about it, his conclusion was that at one time, when the island was uninhabited, it did breed there in some numbers, but that after the island was inhabited, it was gradually exterminated by the frequent robbing of its eggs. This could very easily be done, as the places where it could land and breed were very few, and all on the main island[1] and near the village." The fate of the Stack an Armin bird appears to have been unknown to him, but perhaps the natives responsible for doing it to death as a witch would not care to tell their spiritual adviser of a deed which savoured of heresy.

URIA TROILE, *Common Guillemot.*—Under the name of "Lavy," this species has a place in Martin's history of 1697, and, along with its variable egg, is well described on pages 59 to 61 of his most interesting little book. The Guillemots had quitted their nesting ledges some time before my arrival at St Kilda, but they were sometimes seen in the bay, and they were numerous, along with their young, just off the main island and Boreray on 8th October 1910.

[URIA BRUENNICHI, *Brünnich's Guillemot.*—Milner (p. 2061) tells us that "Brünnich's Guillemot (*Uria bruennichi*) and egg" were taken on Soay on 15th June 1847. Milner himself was not one of the party visiting Soay, and it is impossible to say how far he was responsible for the identification of the bird. The notorious David Graham, of York, was with Mr (afterwards Sir William) Milner and his friends, and most

[1] One of the most suitable breeding haunts for this bird was, in my opinion, the stretch of gradually sloping rocks which form the north-west coast of the island of Dun.

likely it is to him that we are indebted for this and for some other questionable statements.]

Uria grylle, *Black Guillemot.*—There can be no doubt whatever that Martin's description (p. 59 of his book) of the plumage of the bird he names the "Scraber" is applicable to this species, though his account of its habits and nesting refer with equal certainty to the Manx Shearwater. The latter bird is still called Scraber (pronounced Scrapire) by the St Kildans.

The next reference to this bird as a St Kildan species is that of Atkinson, who observed it in 1831. MacGillivray tells us, in 1840, that it was less abundant than the common species; and Milner, in 1847, found it breeding on Dun. The island of Dun and the neighbouring cliffs of Hirta are the main breeding-places and resorts at the present time of this somewhat uncommon species.

We only saw two examples of this bird: an adult off Dun on 4th September 1910, and an immature bird which frequented the head of the East Bay. Here we often watched it flying and swimming under water, in shallows covering the strip of sandy beach, in hot pursuit of the crustacean *Gammarus marinus.*

Alle alle, *Little Auk.*—Dixon (p. 90) was informed that this bird occurs sparingly at St Kilda in winter.

Fratercula arctica, *Puffin.*—Martin (pp. 62-63) treats of this species under the name of "Bouger," and gives an account of the habits of this bird, which, then as now, is one of the species largely captured with rod and noose for food. It is *the* most numerous of the birds resorting to the St Kilda group of islands, where in the summer season it is more abundant than elsewhere in the British Isles.

Only a few birds with late young were seen by us at Hirta, but it was very numerous on the sea around Boreray on 8th October 1910, and on the 12th in 1911.

COLYMBUS GLACIALIS, *Great Northern Diver.*—There is, as yet, only a little information regarding this species in St Kildan seas. Dixon (p. 89) was told that a pair were blown in by a terrible gale in the autumn of 1882. He also informs us (p. 361) that Mr Mackenzie shot an immature specimen, during the first half of June 1885, from several that were swimming in the bay, but failed to secure it ere it had been mutilated by the Great Black-backed Gulls. Steele Elliot (p. 285) observed one in the East Bay on 6th June 1894. Dr Wiglesworth received a specimen which had been picked up dead on the water on 4th December 1903.

We saw an adult bird in the bay on 2nd October 1910.

* PODICIPES AURITUS, *Slavonian Grebe.*—An adult appeared in the bay on 10th October 1911, and was observed busily engaged in diving for small sand-eels, which it captured in the shallow water close in shore.

PROCELLARIA PELAGICA, *Storm Petrel.*—The island of Soay seems to be the main breeding-station of this species, which has been known as a native of St Kilda since 1697, when Martin (p. 63) found it there and described it under the name of "Assilag." Milner (p. 2062) mentions Boreray as a haunt in which he found it in June 1847.

OCEANODROMA LEUCORRHOA, *Fork-tailed Petrel.*—The first British and third known specimen of this bird was obtained at St Kilda by Bullock, who described it in the sale catalogue of his collection as an undescribed Petrel with a forked tail, taken at St Kilda in 1818. At Bullock's sale, in 1819, the specimen was purchased by Dr Leach for the British Museum; and in 1820 it became the type specimen of *Procellaria leachii* of Temminck. Atkinson (p. 224) observed it at St Kilda in 1831, and all subsequent visitors interested in birds have something to say about this species, as the St Kilda archipelago is the main nesting-haunt of this Petrel in the British Isles; indeed it was long thought to be the only one. It is said to breed in all the islands of the group. When Elwes visited St Kilda in 1868 the natives did not discriminate between this species and the Storm Petrel—the rage for collecting birds' eggs had not then set in.

PUFFINUS GRAVIS, *Great Shearwater.*—The late Mr Henry Evans (*Ann. Scot. Nat. Hist.*, 1898, p. 238) records the capture of a Great Shearwater two miles off Dun on 8th August 1897, where, he tells us, "the bird had been obtained before." A party of natives who were fishing observed the bird near their boat and threw some ling's entrails to it, and while eating them it was knocked down with an oar. Two others were killed off the island during the last week of July 1899 (*op. cit.*, 1899, p. 239). Waterston found one dead in the water on 5th July 1905, and another on the 8th.

Many were at sea between Hirta and Boreray on 8th October 1910. When close to Boreray on that date, I saw considerable numbers of these birds engaged in their fine dashing flight around the island, as many as fifty being seen on the wing in company. It was also abundant everywhere at sea, singly and in pairs, between this island and that of Lewis on 8th October 1910, and 12th in 1911.

* PUFFINUS GRISEUS, *Sooty Shearwater.*—One was seen on the sea off the north-east side of St Kilda on 12th September 1910. On 8th October several came under our notice off Boreray; and others, thirteen in all, between the St Kildan islands and the Flannan groups—single birds, except a party of three.

A bird of this species was seen off Boreray on 12th October 1911, and several others during the voyage to Lewis. It was one of the species which contested with the Gannets, Gulls, Pomatorhine Skuas, Greater Shearwaters, and Fulmars for the fish-offal cast out by the cook of the trawler. In its eagerness to obtain this food it plunged into the water, completely immersing itself.

This species does not appear to have been hitherto recorded for Hebridean seas.

PUFFINUS ANGLORUM, *Manx Shearwater.*—This is in part the "Scraber" of Martin (see Black Guillemot), who well describes its habits and affords some curious information regarding them. Dixon (p. 94) says that as many as four hundred have been captured on Soay in a single night. Wiglesworth (p. 63) mentions that it is still plentiful, but less so than formerly—a fact which the natives attribute to its having been driven out by the increase of the Puffin. Neil Ferguson tells me that it breeds on Hirta, Soay, and Dun, but not on Boreray.

It is a summer visitor to the isles, and we only saw one—namely, at sea when approaching the archipelago, on 1st September 1910.

FULMARUS GLACIALIS, *Fulmar*.—This is, and always has been, the most important bird from the St Kildan's point of view, for it furnishes them with the staple item of their bill of fare. About 9600 young Fumlars, just ready to take wing, and about six weeks old, had been captured and salted down just prior to our visits. Next to the Great Auk, it formerly held the premier place among the St Kildan birds in the eye of the ornithologist, for until some quarter of a century, or a little more, ago, this group of islands afforded the only nesting-haunts of this species in the British seas. Martin (pp. 55-58) gives a quaint account of this bird and the method adopted for its capture in 1697.

A very few young birds were still nestling on the ledges of the great cliffs on our arrival, but soon after they were all on the wing. Thousands of old birds were to be seen sailing along the faces of the breeding cliffs down to the middle of September; they were far less numerous on the 25th, and all had gone out to sea by the end of the month. Their object in leaving the islands at this season is probably for the purpose of moulting, for they return after a month's absence, and remain all the winter in the immediate neighbourhood.

We found these birds were inclined to be curious, for they often flew within a foot or two of our heads when we made an appearance in the vicinity of their haunts.

There were numbers at sea between St Kilda and the Flannan Isles in October 1910 and 1911.

The northern form, known to the St Kildans as the "Blue Fulmar," has several times been captured, but is rare. It was described to me as being "blue" all over, including the bill. Blue should probably be rendered as grey. An albino specimen, a young bird, was captured in August 1910, and is now in the Royal Scottish Museum.

Addendum

*Corvus monedula, *Jackdaw.*—I received one which had been captured on 9th December 1911.

BIBLIOGRAPHY

The following works have been more or less frequently referred to in the foregoing accounts of the various species. They are given in the chronological order of their publication:—

1698. Martin, Martin.—*A late Voyage to St Kilda, the remotest of all the Hebrides or Western Isles of Scotland:* London, MDCXCVIII.

[This book will always remain a St Kildan classic. Martin arrived at St Kilda on the 1st of June 1697, and he remained three weeks on the island. He mentions or describes twenty-one species of birds.]

1764. Macaulay, Rev. Kenneth.—*The History of St Kilda:* London, MDCCLXIV.

[Macaulay visited St Kilda in June 1758. Chapter viii. (pp. 131-161) is devoted to an account "Of the Sea and Land Fowls at Hirta," and adds several species to Martin's list.]

1832. Atkinson, G. C.—"A Notice of the Island of St Kilda, on the North-west Coast of Scotland," *Trans. Nat. Hist. Soc. Northumberland, Durham, and Newcastle-on-Tyne*, vol. ii., pt. i., pp. 215-225 (1832).

[An account of a visit made on 31st May 1831, including a list at p. 224 of the birds observed.]

1840. MacGillivray, John.—"An Account of the Island of St Kilda, chiefly with reference to its Natural History, from Notes made during a visit in July 1840," *Edinburgh New Phil. Journ.*, vol. xxxii., pp. 47-70.

[Contains interesting and valuable matter.]

1842. WILSON, JAMES.—*Voyage round the Coasts of Scotland and the Isles*, 2 vols., London, 1842.

[Pages 1-113 of vol. ii. give an excellent account of a visit paid to St Kilda on 2nd and 3rd August 1841. Forty species of birds are mentioned.]

1848. MILNER, W. M. E.—"Some Account of the People of St Kilda, and of the Birds of the Outer Hebrides," *Zoologist*, 1848, pp. 2054-2062.

[Mr Milner, afterwards Sir Wm. Milner, Bart., spent three days—14th to 16th June 1847—at the islands of the St Kilda group.]

1885. DIXON, CHARLES.—"The Ornithology of St Kilda," *The Ibis*, 1885, pp. 69-97 and 358-362.

[Mr Dixon spent nearly a fortnight there in June 1884. A complete list of the then known avifauna of the group is given.]

1895. ELLIOT, J. STEELE.—"Observations on the Fauna of St Kilda," *Zoologist*, 1895, pp. 281-286.

[The result of three weeks spent on the island in June 1894.]

1903. WIGLESWORTH, J.—*St Kilda and its Birds* : Liverpool, 1903.

[Dr Wiglesworth passed three weeks at St Kilda in June 1902.]

1905. MACKENZIE, Rev. J. B.—"Notes on the Birds of St Kilda. Compiled from the Memoranda of the Rev. Neil Mackenzie" [1829-1843], *Ann. Scot. Nat. Hist.*, 1905, pp. 75-80 and 141-153.

[These memoranda relate to the years 1829-1843.]

1905. WATERSTON, Rev. JAMES, M.A.—"Notes on the Mice and Birds of St Kilda," *Ann. Scot. Nat. Hist.*, 1905, pp. 199-202.

CHAPTER XXIV

THE FLANNAN ISLES AND THEIR BIRD-VISITORS: ANOTHER STUDY OF THE FAR-WESTERN PASSAGE MOVEMENTS

Away to the westward of Lewis, the northernmost isle of the Hebrides, and hence right out in the Atlantic, lies an insignificant archipelago of islets, known only to the few who are interested in the remotest spots amid the British seas—namely, the Flannan Islands.

Here, in the year 1899, a lighthouse was erected on the largest of this group, Eilean Mor. The keepers at this lonely station were, in due course, requested to keep a record of the visits of any migratory birds that came under their notice. This request they most obligingly complied with, and the records obtained were of a somewhat surprising nature, since they unmistakably indicated that the islands, in spite of their far-western situation, were annually visited by a considerable number of migrants representing not a few species. The unexpected success which attended these investigations, suggested that it was desirable that a visit should be paid to the island during the period of the autumn movements, in the hope of adding to our knowledge of its migratory bird-life, and of stimulating and interesting the light-keepers in the work of observing in the future. With these objects in view, Mr T. G. Laidlaw and I spent sixteen days

PLATE XX.

[*Photo: C. Dick Peddie.*

FLANNAN ISLANDS: EILEAN MOR FROM THE EAST.

there in the autumn of 1904, during which period we resided in the lighthouse, by permission graciously granted by the Commissioners for Northern Lighthouses.

The Flannan Islands, known also as The Seven Hunters, form a small group of uninhabited isles, lying from 20 to 23 miles west of Gallan Head, Island of Lewis, and 40 miles north-east of St Kilda, with the exception of which they are among the most western islands of Great Britain.

They may be said to form two groups—an eastern one, comprising Eilean Mor, Eilean Tigh, Soraidh, and one that is nameless; and the western one, composed of Rhoderheim, Bronna Cleit, and Eilean Gobba. In addition there are several skerries and rocks. The seven main islands are precipitous, rising on all sides direct from the ocean; and this fact, together with their exposed situation, renders them extremely difficult to land upon. There are probably no wilder spots to be found in the British seas.

The lighthouse is situated on Eilean Mor, the most northerly of the eastern group. This isle is entirely girt by a belt of cliff, highest on the north and east, where it attains to a height of 282 feet, and lowest and mostly under 200 feet on the south. Thus the top of the island is an elevated plateau, sloping towards the south, and has an area of 16 acres, clothed with grass and sea-pink, and with here and there bare patches of peaty turf and exposed rock (Lewisian gneiss). Where not actually precipices, the banks facing the sea are extremely steep, indeed all but perpendicular, and are clothed with a profuse growth of herbage, chief among which is a marguerite (*Chrysanthemum inodorum*). These dangerous slopes are much resorted to as feeding-grounds by the smaller species of migratory birds, for amid the shelter afforded, they are comparatively safe from the assiduous attentions of an ever-present Peregrine, and the occasional raids of the Merlin. There is no cover elsewhere, unless some short, narrow runnels, connecting a series of diminutive pools, on the exposed plateau, can be regarded as such.

There are three ancient dwellings on the island. The chief of these resembles a large dog-kennel, has a low entrance, is supposed to have been built for devotional purposes, and is named on the Ordnance Map the "Blessing House." Far more interesting are two brochs, in an excellent state of preservation, each with a couple of square chambers, and surmounted with a beehive-shaped roof.

The islands are resorted to annually as breeding-stations, by hosts of sea-birds, the foremost among which, in point of numbers, is the Puffin; and the most interesting, the Fork-tailed Petrel and the Fulmar.

Ornithologically, however, the islands are chiefly remarkable for the streams of birds which visit them in spring when on passage to their northern summer homes, and in autumn when returning to their southern winter retreats. These movements are surprising for the number of species and individuals visiting Eilean Mor, when its situation, its very limited area, and its singularly few attractions for migrants are taken into consideration.

The data amassed regarding these passage movements cover a period of no less than twelve years, and afford valuable information on the dates between which

PLATE XXI.

[*Photo: C. Dick Peddie.*

FLANNAN ISLANDS: EILEAN MOR FROM THE SOUTH.

[*To face p.* 252.

they are performed by many species during both spring and autumn.

In addition to regular migrants, these visitors include a number of extremely rare species, such as the Short-toed Lark, the Two-barred Crossbill, the Siberian Skylark, and the Pratincole. For such as these, the Flannan Islands are strange destinations; they are among the very last places at which one would expect them to appear.

In winter these islands are visited during periods of great severity and snow by a number of birds seeking genial retreats. These evicted birds, however, do not, as a rule, remain long on the island.

We arrived at Eilean Mor, after a rather rough passage, on the morning of 6th September, and remained in residence at the lighthouse until the next "relief" was effected on the 21st.

The result of our investigations added considerably to the knowledge we hitherto possessed on the migrants visiting the archipelago. We felt, however, that we might have accomplished more, if only we could have visited daily the island, Eilean Tigh, lying immediately to the south, and separated from us by a narrow sound. It would not only have afforded an important extension of our field for observation, but also additional scope for our energies, which were decidedly circumscribed.

The lantern of the lighthouse is of great power, 140,000 candle, and is concentrated into three brilliant slowly revolving beams. Few birds, however, visited it during our sojourn on the island, which was not surprising, for the weather conditions were almost uniformly unfavourable for "a night at the lantern," being clear and free from moisture.

We were gratified to find two of the lightkeepers, Messrs Begg and Anderson, keenly interested in our researches, and since our departure they have furnished excellent records, in many cases accompanied by birds killed at the lantern. On Mr Begg's leaving the station, Mr Anderson has continued the observations down to 1910. Both these observers have earned our gratitude for their very valuable contributions. The total known avifauna now stands at 115 species—a very remarkable figure, but there are, doubtless, additions to come, for, so far, every year has added its quota.

The following are the species which came under notice during our residence on the island:—

RAVEN, *Corvus corax.*—A pair is resident among the islands, remaining all the year round, and nesting on one of the outer group. During our visit this couple was seen daily, but they had evidently dismissed their young before we arrived. One day a strange pair appeared, to the evident concern of the titular proprietors, who were very unsettled and noisy during the trespass upon their domain.

GREY CROW, *Corvus cornix.*—A pair seen occasionally at all seasons, but said not to breed; at any rate their young have not been observed. These birds were seen throughout our visit, and one of them was shot, but the survivor had a fresh mate two days later.

SNOW-BUNTING, *Plectrophenax nivalis.*—A winter visitor in numbers. The earliest record of its appearance came under our notice, two being seen on 14th September 1904. A few usually arrive later in the month, but the great flights do not appear until October, when they are sometimes noted as in "thousands." A considerable

[Photo: C. Dick Peddie.

FLANNAN ISLANDS: THE EAST LANDING-PLACE.

number stay the winter. Late in March great flocks gather on the island previous to departure, and many are then in summer plumage. It is also observed in numbers on passage, with other emigrants bound north in April.

LAPLAND BUNTING, *Calcarius lapponicus.*—We found this species present on our arrival on the island on 6th September, and the keepers told us that the birds were seen by them on the 3rd, and were thought to be Snow-Buntings. It was one of the first birds seen by us, and was still present on the island at the date of our departure. From careful computations we set down their number at from thirty to forty individuals, but there may have been more, for we were not able to visit the other islands, some of which were equally well suited for their requirements. They sought food on the bare patches of peat, and down the face of the cliffs; and at night roosted among the coarse grass growing on the top of the island and on the steep banks. When on the wing they were inclined to be noisy, some of their notes resembling those of a Linnet, others those of the Snow-Bunting, especially its call note *tūke, tūke.* They were usually seen in small parties, perhaps families, and the examples obtained were adults of both sexes and young birds of the year.

SKYLARK, *Alauda arvensis.*—Strange to say, we only saw a single example—namely, an immature bird, on 12th September. The northern migrants do not arrive until October, and the passage southwards lasts until the end of November.

SHORT-TOED LARK, *Calandrella brachydactyla.*—This unlooked-for visitor from Southern Europe is believed to have appeared on the night of 19th September. It was detected as a stranger early the following day, its light colour and small size attracting attention. Its note on rising on the wing, when disturbed, resembled that of a Skylark. It proved to be a female, and is the first recorded occurrence of this species for Scotland, and in 1904 the Flannans were the most northerly known locality in Europe visited by this bird.

WHITE WAGTAIL, *Motacilla alba.*—This species is probably common on migration in both spring and autumn. It was certainly one of the most abundant migrants observed by us from the day of our arrival, 6th September, to our departure on the 21st. During this period two distinct arrivals took place—namely, on the night of the 8th, along with Meadow-Pipits and Wheatears, and again on the 11th, when it appeared along with the same species at 2.30 A.M., at which hour several came to the lantern. The numbers dwindled after the 13th, but a few were present down to the 23rd, as is testified to by specimens received after our departure.

[GREY-HEADED WAGTAIL, *Motacilla thunbergi.*—A bird, believed to be of this species, appeared on the morning of 20th September, and for two days baffled all our attempts to solve its identity with certainty. It was shy, wary, and restless in the extreme, and never allowed us to get a good view of it through our binoculars.]

MEADOW-PIPIT, *Anthus pratensis.*—This species was abundant during the whole period covered by our visit; and there were considerable arrivals on 8th September, with White Wagtails and Wheatears; on the 10th, from 8.30 to 10 P.M.; and again on the 11th, with Wheatears and White Wagtails.

ROCK-PIPIT, *Anthus obscurus.* — Resident and extremely abundant. It was marvellously tame, being

[*Photo: C. Dick Peddie.*

FLANNAN ISLANDS: THE SOUTH LANDING-PLACE.

almost indifferent to one's presence, and came freely into the house. Some examples were much larger than others, and it is probable that all were not natives of the island.

GARDEN-WARBLER, *Sylvia borin.*—One was observed searching for flies on the face of a crevasse on the north cliff on 16th September, and remained until the 18th, perhaps longer.

WHEATEAR, *Saxicola œnanthe.*—Was very abundant on passage southwards during the whole of our visit. There were considerable arrivals on the night of 8th September, along with Pipits and White Wagtails; and again along with the same species at 2.30 A.M. on the 11th, when it was observed at the lantern, and was in swarms all over the island during the day. There was another arrival on the 17th, when it appeared at the lantern from 11 P.M. onwards. All the birds observed were, with one exception, in the russet plumage of autumn and winter; the exception was a male, which still displayed traces of summer plumage. About four pairs of Wheatears nest annually on the island.

PIED FLYCATCHER, *Muscicapa atricapilla.*—Was first observed on 13th September, when three were seen busily engaged capturing insects on the face of rocks in a rift in the cliffs on the north side of the island—as wild a spot as it is possible to imagine. Here these birds were seen daily until the 20th, and perhaps remained beyond that date. All were in the grey dress of autumn. This species was an addition to the avifauna of the Outer Hebrides.

SWIFT, *Cypselus apus.*—The Swift is regarded as a rare visitor. Previous to our visit, there were three records of its occurrence—two for June and one for July.

On 16th September I saw one at 10 A.M., during heavy rain and a strong south-east breeze. And on the following day another was flying under the north cliff for several hours, seeking shelter from the strong southerly breeze.

PEREGRINE, *Falco peregrinus*.—A pair of Peregrines nest annually on one of the outer islands of the group. During the early days of our visit we saw the old birds accompanied by one of their offspring, a male, which they tried their best to drive away from the islands, but all their bullying failed. The old birds left about 14th September, but the young male remained, and was a scourge to all the small migratory birds resorting to Eilean Mor. It used to dash many times a day over the exposed plateau in search of prey, making sad havoc in the ranks of the travellers. All our efforts to put an end to its ravenings were futile, owing to the impetuosity of its dash and the suddenness of its appearance.

MERLIN, *Falco æsalon*.—A pair appeared on the morning of 11th September. They only remained for the day, during which the small migrants had a bad time of it. We had a captive Fork-tailed Petrel, which had been taken at the lantern the previous night. This we released in order to observe its manner of flight, no Merlins being in sight at the time; no sooner, however, was the Petrel well on the wing, than one of the Merlins appeared in hot pursuit, but the Petrel, we believe, reached the sea in safety.

SHAG, *Phalacrocorax graculus*.—Very common all the year round, and breeding in considerable numbers. We saw old birds feeding young as big as themselves. The importunities of these fledglings were of a very persistent nature, and not at all appreciated by the parent birds, who did their best to shake off their lazy offspring.

GANNET, *Sula bassana.*—Although observed all the year round, Mr Begg informs me that only a few are seen during the winter months. Soon after the middle of March flocks are observed constantly passing towards St Kilda. During our visit it was always common, but only adult birds were seen.

HERON, *Ardea cinerea.*—An occasional visitor. We saw five of these birds—one on 10th September, two on the 11th, and one on 17th and 18th—all of them young. They did not stay long, which was not surprising, for the islands afford no possible feeding-grounds.

BRENT GOOSE, *Branta bernicla.*—On 16th September we saw a bird of this species flying to the south-east.

EIDER DUCK, *Somateria mollissima.*—Nests commonly on some of the islands, but is not seen during winter. Old and young were very numerous throughout our visit. They usually formed a straggling flock, and fed under the lee of the islands. Several females, accompanied by half-grown young ones, kept apart from the main body.

CORN-CRAKE, *Crex crex.* — A casual visitor. One appeared on 15th September.

RINGED PLOVER, *Ægialitis hiaticola.*—On 12th September, a single bird in immature plumage appeared, and was followed by another and similar bird on the 19th. The two frequented the bare patches among the sea-pink on the top of the island down to the date of our departure.

GOLDEN PLOVER, *Charadrius pluvialis.*—Occurs on passage, in both spring and autumn. The first visitors during the autumn of 1904 were four birds which appeared on 12th September. These were followed by a few others on the 13th and 17th.

LAPWING, *Vanellus vanellus.*—Occurs regularly on passage in both spring and autumn. We had only one visit of Lapwings during our stay. On 13th September three appeared in the early morning, and remained a few hours on the island.

TURNSTONE, *Strepsilas interpres.*—Occurs occasionally during the autumn passage and in winter; but the island, with its precipitous sides, affords little attraction for this bird, and it does not remain long. We observed a single bird on 7th September; and two appeared on the 10th and remained for several days, during which one fell a victim to the Peregrine Falcon. These birds, like the Ringed Plovers, frequented the bare patches amid the sea-pink on the top of the island.

OYSTER-CATCHER, *Hæmatopus ostralegus.*—A common summer visitor to the islands. Several pairs nest on Eilean Mor, and leave along with their young in August. We saw only a single example during our visit—namely, on 14th September.

COMMON SNIPE, *Gallinago gallinago.*—A frequent visitor, though usually appearing in small numbers, during the spring and autumn migratory movements. A few were observed during our visit, the first of which appeared on 14th September. It also occurs at intervals in October and November.

JACK SNIPE, *Gallinago gallinula.*—The chief migratory movement witnessed during our visit was a very remarkable one on the part of this species. The first immigrant was observed by us on the evening of 14th September, and this was followed on the night of the 16th or the early hours of the 17th by a great rush.

The morning of the 17th was characterised by a high wind from the south-west, accompanied by a downpour of rain. On venturing out soon after 8 A.M., I found the island swarming with Jack Snipe. They were in astonishing numbers, and sheltering behind rocks, tufts of grass, in the small pools and runnels, and even down the face of cliffs on the north-side. In walking across the island I put up a continuous stream of them, in spite of the fact that the birds sat like stones, and only those arose on the wing which lay directly in my course, and when I was close upon them. It was a most remarkable experience, and one entirely unexpected, in view of the species and the locality. It was surprising, too, to find that the Jack Snipe migrated in such vast packs. I believe that a record "bag" could have been made on this 16-acre island, in an hour's shooting. I have no doubt that they were abundant on the other islands of the group, and probably especially so on the adjacent and comparatively flat-topped Eilean Tigh. The birds were present in numbers the entire day, but nearly all, perhaps all, departed during the night, and the few (eight) seen on the following day may have been fresh arrivals, as may also those observed on the 19th and 20th.

DUNLIN, *Tringa alpina.*—The islands afford no suitable feeding-grounds for this bird, but a few visit them during the seasons of passage. Two of these migrant visitors came under our notice—namely, one on 8th September, and another on the 16th.

CURLEW, *Numenius arquata.*—A regular visitor on passage in both spring and autumn. A few small parties came under our notice on 11th, 12th, and 13th September.

WHIMBREL, *Numenius phæopus.*—It would seem probable that before the lighthouse buildings were commenced, this bird was a summer visitor to Eilean Mor, for during the first summer the artisans were engaged there, a nest was found near the ancient building, known as the "Blessing House," and the eggs taken for food. We saw only two birds of this species—namely, on 15th September.

GREATER BLACK-BACKED GULL, *Larus marinus.*—About two pairs are resident on Eilean Mor, and nest there; and a few are also to be found on the other islands during the summer months. During our stay both old and young were seen daily in some numbers.

LESSER BLACK-BACKED GULL, *Larus fuscus.*—Does not nest on the main island, but is often seen during the summer months, for there is a colony on the adjoining Eilean Tigh. This species had practically left the islands before our visit, for only one adult and a few young birds came under notice.

HERRING-GULL, *Larus argentatus.*—About six pairs only breed on Eilean Mor, but many resort to the other and more secluded islands of the group. A party consisting of about a score of birds was present during our sojourn, and remained on Eilean Mor throughout the winter, being attracted by the refuse thrown out from the lighthouse.

KITTIWAKE, *Rissa tridactyla.*—The commonest gull during the nesting season. They evidently depart early in the autumn, for we did not observe a single adult example during our visit, but a few in first plumage were present.

RAZORBILL, *Alca torda.*—Thousands of these birds haunt the cliffs during the nesting season. They arrive late in February or in March. We did not see the Razorbill in the vicinity of the islands: all had departed before our arrival on 6th September.

GUILLEMOT, *Uria troile.*—Vast numbers make their homes on the cliffs during spring and summer, arriving from the middle to the end of February. They had quitted their breeding-haunts before our visit, and we only saw old birds, accompanied by their full-grown young, at some distance from the islands, on the occasions of our arrival and departure.

BLACK GUILLEMOT, *Uria grylle.*—This characteristic west-coast species thins out in the Hebrides, and we could not obtain any evidence that it breeds on any of the Flannan Islands. It is possible, however, that a few do, and that they have escaped notice among the vast crowds of rock-birds present during the summer. We observed odd birds, sometimes a pair, close to the base of the cliffs of Eilean Mor on several occasions.

PUFFIN, *Fratercula arctica.*—The commonest summer visitor to Eilean Mor, and probably to the other islands of the group. On the top of the main island, where the turf is suitable for the formation of their burrows, there are colonies of thousands, and vast numbers resort to the holes and crevices on the face of the cliffs. The great breeding crowd had left ere we arrived, and the very few that remained were detained by their unfledged young, some of which were still half-clad in down. The industry of this comical bird is marvellous. We sometimes sat near their burrows, and the constancy with which the old birds arrived with strings of small fish hanging from their bills, was quite remarkable; they seemed to be coming in every few minutes, and the young must surely be very voracious little creatures.

Slavonian Grebe, *Podicipes auritus.*—We observed a bird of the year on 16th September, off the east landing place. A strong southerly breeze prevailed at the time, and the bird found there a convenient harbour, and passed the entire day in diving and sleeping. It often came quite close to the face of the rocks from which we were watching it.

Storm-Petrel, *Procellaria pelagica.*—Very numerous during the summer, when they fly noisily about the island during the night-time. They breed on Eilean Mor, and probably on the other islands, in abundance. Many chicks, some of them quite recently hatched, were found during our visit in September, and we saw young ones in every stage, from a few hours old (tiny balls of pretty lavender-grey down) to birds full-grown and fully feathered, except that they had a bunch of down still present on the abdomen. The old birds were entirely absent during the daytime, leaving even the tiny chicks to take care of themselves, and did not return until darkness set in, when they tend their young, and depart again in the early morning, probably to spend the day far out at sea in search of food. We opened out a number of their nesting-holes at all hours of the day, but the old birds were always absent, except in one instance where the young had only just emerged from the egg. They nest in the remains of the old buildings, in holes in turf, and under stones among grass. The nest is a mere mat composed of dry roots, grass, etc. I received a young one in full down, which had been taken on 3rd October; probably the first egg of this pair had been taken or destroyed. Occasionally they visited the lantern.

Fork-tailed Petrel, *Oceanodroma leucorrhoa.*—

The Flannan Isles may be regarded as one of the chief breeding-stations of this species in the British Isles—and will probably long remain so, thanks to the inaccessibility of these islands. On Eilean Mor they are abundant, more so than the Storm-Petrel, and, like that species, fly noisily over the island during the short summer nights. They lay earlier than *pelagica*, the earliest date for their eggs being 29th May, but their nesting habits are very similar. We found their nurseries under stones among turf; in holes in turf, overgrown with grass, yet showing not the slightest signs of the incomings or outgoings of the occupants; and in the walls of the old buildings. In some of the burrows, the mat-like cradles of roots and fibrous vegetable matter were placed several feet from the entrance. The chicks are much darker in colour than those of the Storm-Petrel, being sooty black; they were also, as a rule, a little more advanced, but youngsters only a few days old were found during the early days of our visit, as well as others in every stage up to those almost ready to fly. The old birds were entirely absent during the daytime, and only occasionally came under notice at night when they visited the lantern.

FULMAR, *Fulmarus glacialis*.—A few pairs have bred on the outer islands for several years, and in 1904 two couples bred on Eilean Mor for the first time. We saw this bird on several occasions during our visit.

GREAT SHEARWATER, *Puffinus gravis*.—On leaving the islands on 21st September, we saw, from the bridge of the "Pole Star," a few of these birds at sea a little distance off the Flannans.

The following is a short account of all the *migratory*

birds which are known to have occurred at Eilean Mor during the past twelve years (1900-1911); including some of the dates upon which they have appeared.

ROOK, *Corvus frugilegus.*—The nature of the visits of this bird is somewhat uncertain. It appears annually in small numbers during March (14th earliest) and April (10th latest), and some years in May (as late as the 24th). Some of these visits may possibly relate to passage movements. In autumn it has only appeared on two occasions, on 24th and 26th October 1907.

JACKDAW, *Corvus monedula.*—There are two records only for the visits of this species—namely, of single birds on 22nd February 1901, and 7th November 1905.

STARLING, *Sturnus vulgaris.*—A regular visitor during the periods of the spring and autumn passages to and from the north, and in winter during severe weather.

The spring movements date from 28th February to 17th April, March being the main month for their performance.

The return passage southwards is sometimes recorded for late September (27th earliest), and is in progress until mid-November. In October the bird often occurs in great numbers along with other migrants.

CHAFFINCH, *Fringilla cœlebs.*—Appears annually, sometimes in considerable numbers, in autumn; but is neither frequent nor numerous in spring. It is also an occasional visitor in winter.

The autumn observations date from 4th September to mid-November, but it does not become numerous until mid-October.

The earliest record in spring is for 6th March, and the latest for 17th May.

BRAMBLING, *Fringilla montifringilla.* — Has been recorded in a few instances for spring and autumn. There is only one entry in the schedules for spring—namely, of a single bird on 7th May 1908.

The autumn visits date from 1st October to 3rd November, and chiefly relate to one or two birds; but on 18th October 1908, several appeared, along with a number of other species.

GREENFINCH, *Chloris chloris.*—Occurs in autumn, winter, and spring. The earliest date for its appearance at the first season is chronicled for 27th October. The winter and spring occurrences merge, for it has been recorded for November, December, January, February, and March (10th latest). Occasionally large flocks arrive late in October, but usually the data relate to parties of several to a dozen birds.

[SISKIN, *Spinus spinus.*—There is only a single record for this species, and that is a somewhat doubtful one. A single bird is said to have occurred on 24th April 1901.]

LINNET, *Acanthis cannabina.* — Has occasionally been recorded for February, March, July, and October, four being the greatest number observed.

TWITE, *Acanthis flavirostris.*—A summer visitor to the islands. Two pairs breed on Eilean Mor, arriving in spring, and departing early in the autumn. We did not see the bird in September 1904.

MEALY REDPOLL, *Acanthis linaria.*—A small party visited the island during September 1907, when birds were first observed on the 7th, and remained until the 16th. A specimen captured was sent for identification.

GREATER REDPOLL, *Acanthis rostrata* (*Acanthis linaria rostrata*).—Several of this large Greenland race of Redpoll visited Eilean Mor in October and November 1905, and specimens were forwarded to me. This bird was very abundant at Fair Isle the same season.

Crossbill, *Loxia curvirostra.* — A considerable number visited the islands during the memorable irruption of the summer of 1909. They arrived late in June, remained in some numbers until the second week of August, and the last was observed on 22nd September.

Two-barred Crossbill, *Loxia bifasciata.*—An adult male was shot on 21st July from among the common Crossbills which frequented the island during the summer of 1909. In 1910 a male was captured on 14th August.

Yellow Bunting, *Emberiza citrinella.*—There are three records for the occurrence of this species: one on 17th September 1900; twelve on 30th October 1905; and ten on 12th November 1905.

Reed-Bunting, *Emberiza schœniclus.*—There is a single record only—namely, of a female on 19th May 1906, which was sent for identification.

Lapland Bunting, *Calcarius lapponicus.* — Since our discovery of this species, Eilean Mor has been visited annually. The dates of the appearances of these supposed rare visitors are interesting on account of their earliness and regularity — namely, from 3rd to 11th September. Specimens as vouchers have been sent.

Snow-Bunting, *Plectrophenax nivalis.*—Appears at the islands on passage in the spring and autumn, and also as a winter visitor.

Many are observed in spring from mid-March and during April, fewer in May, down to the 19th. The latest date for spring stragglers of the rearguard is 4th June.

The earliest record for autumn is for 3rd September.

Great numbers arrive, sometimes thousands, in October and November. Some are seen all winter.

WHITE-THROATED SPARROW, *Zonotrichia albicollis.*—An adult male of this American bunting was shot near the lighthouse on 18th May 1909. (*Ann. Scot. Nat. Hist.*, 1909, p. 246.) The date is, perhaps, not in favour of its having arrived in the British Islands unassisted.

SKYLARK, *Alauda arvensis.*—Occurs regularly on passage in spring and autumn, usually in small numbers. A few visit the island during severe weather in winter, but do not remain long.

From the records it appears to be most numerous in spring, when it occurs from mid-February, during March, and to mid-April (15th latest date).

In autumn a few have been seen in September, but it is not until October and during November that the southern passage of limited numbers is observed.

ASIATIC SKYLARK, *Alauda cinerea* (*Alauda arvensis cinerea*).—An example of this Asiatic race of the skylark was killed at the lantern on 24th February 1906. This form was described by Ehmcke in the *Journ. für Ornithologie* in 1903 (p. 149), as *Alauda cinerea*, and is the *Alauda arvensis cinerea* of Dr Hartert's *Vögel der paläarktischen Fauna* (p. 247). This capture well illustrates the advantage that accrues from a knowledge of racial forms, since it has enabled me to determine whence came this remarkably grey skylark to our shores—a bird that has not hitherto been detected in Europe, except in the Far East. According to Dr Hartert (*loc. cit.*), it has its home in Western Siberia, Turkestan, Persia, and possibly in Palestine; and in winter is found on the northern side of the Caucasus, in Egypt, Tunis, and Algeria. I have to thank Mr Rothschild and Dr Hartert for facilities which rendered its determination possible — namely, by a comparison with specimens in the unrivalled collection of Palæarctic Birds in the Tring Museum. This interesting specimen has been most kindly presented to the collection of birds in the Royal Scottish Museum by Mr George Girdwood of Dumbarton.

SHORT-TOED LARK, *Calandrella brachydactyla.*—The visit and capture of this interesting waif, on 20th September 1904, has already been described on pp. 255-6.

PIED WAGTAIL, *Motacilla lugubris.*—Appears to be an uncommon visitor, and is, no doubt, in some cases not discriminated from the next species.

It is recorded for April (20th earliest) and May (19th latest). An adult male, obtained during the former month, was sent to me.

In autumn the references to its appearance are from August (11th earliest) to September (15th latest).

WHITE WAGTAIL, *Motacilla alba.*—A regular bird of passage in the spring and autumn, and one of the commonest migrants, being observed in large parties at both seasons.

The spring records range from 2nd May to 3rd June; and those for autumn, from 10th August to 8th October, being most numerous during September.

[GREY-HEADED WAGTAIL, *Motacilla thunbergi* (= *borealis*).—The appearance of this species on 20th September 1904 has already been fully treated of (p. 256).]

MEADOW-PIPIT, *Anthus pratensis.*—Though one of the commonest visitors on passage during the autumn, this bird seems to have practically escaped notice, at least in the schedules, for spring.

There is only one record for spring—namely, 20th

May 1904, when it occurred at the lantern in numbers, along with Wheatears.

Its autumn visits are frequent, and are recorded from 20th August to 8th October.

TREE-PIPIT, *Anthus trivialis.*—A single visit only is known. One killed at the lantern on the night of 25th September 1908 was sent to me. This species had not previously been known to occur in any of the islands of the Outer Hebrides.

ROCK-PIPIT, *Anthus obscurus.*—Resident all the year round, and very numerous. Many perish in winter, probably from want of food, but the lightkeepers say from exposure.

There are entries in the records of their presence in great numbers on 21st April 1901, and on 17th September 1902, and these may refer to passage movements on the part of this species.

RED-BACKED SHRIKE, *Lanius collurio.*—A bird of the year was killed at the lantern on 14th September 1909, and was forwarded for identification. This species is not known to have occurred previously in the Outer Hebrides.

GOLDCREST, *Regulus regulus.*—Several occurred in a rush of migrants on 30th October 1908. This is the only known visit of this species to the island, and bears out the general pronouncement on the scarcity of the Goldcrest as a migrant in the Outer Hebrides.

COMMON WHITETHROAT, *Sylvia sylvia.*—One captured in May 1907 was sent for identification. There are records for 3rd June and 23rd September 1902, but these are of a somewhat doubtful nature.

LESSER WHITETHROAT, *Sylvia curruca.*—One was obtained on 23rd September 1904, and sent to me.

Blackcap, *Sylvia atricapilla.*—One killed during a rush of migrants on 29th October 1908, was sent to me. There is a doubtful record, 8th March 1904.

Garden-Warbler, *Sylvia borin.*—The bird seen from 16th to 18th September 1904, as already related (p. 257), is the only known instance of the visit of this bird, and the second one for the Outer Hebrides.

Willow-Warbler, *Phylloscopus trochilus.* — There are a few records of the visits of single examples of this bird for both spring and autumn. In spring one was obtained early in May 1909; one on 14th May 1906; and one on 3rd June 1906; all of which were sent.

The autumn visits were for 2nd August 1904, 30th September 1907, and 19th October 1907.

Sedge-Warbler, *Acrocephalus schœnobænus.* — Has twice been known to visit the island—namely, on 16th June 1905 and early in May 1907. Both the birds were sent for determination.

Song-Thrush, *Turdus musicus.*—Occurs on passage in both spring and autumn, and occasionally as a visitor during severe weather in winter.

Has been observed in spring as early as 20th February, and on passage during March and on 1st April.

In autumn has occurred from 1st October to 19th November, often in great numbers with other migrants.

Redwing, *Turdus iliacus.*—A regular visitor in numbers during both seasons of passage; and an occasional visitor during severe weather, especially snow, in winter.

There are occurrences in the latter half of February and during March, of many birds in April, and of stragglers down to 14th May.

The autumn movements range from 2nd October to 19th November, when the bird sometimes appears in thousands.

On 5th December 1902, numbers appeared with Fieldfares; many of them were found afterwards too weak to fly.

Fieldfare, *Turdus pilaris.*—A bird of double passage, and a visitor during stressful periods in winter.

In spring, has been observed from 15th April to the end of the first week of May, and occasionally as late as 6th June.

The earliest record for its appearance in autumn is 3rd October, and it occurs in numbers from the middle of that month down to 19th November.

Blackbird, *Turdus merula.*—Regular and numerous on both passages, and a visitor during cold spells in winter.

In spring it has been observed from the earliest days of March to 12th May; and in autumn from 29th September to 20th November, often in considerable numbers.

Ring-Ouzel, *Turdus torquatus.*—This species has visited the island in a few instances only. In the spring of 1908 single birds were observed on 28th and 29th April; and in 1910 one appeared on 14th May. On 2nd October 1907, several appeared, and on the two following days single birds were seen.

Redbreast, *Erithacus rubecula.*—Occasionally occurs, chiefly in small numbers, in spring, autumn, and winter. The spring visits date from 25th March to 13th May; those of autumn from 28th October to 7th November; and the winter appearances from 30th November to 12th February.

Redstart, *Ruticilla phœnicurus.*—Single examples have been obtained on five occasions in the autumn, on dates ranging from 23rd September to 7th October. Several of these were killed at the lantern and were forwarded to me.

Black Redstart, *Ruticilla titys.*—This somewhat uncommon visitor to Scotland has occurred singly on five occasions, and its visits on three of these have been vouched for by specimens. Four of the occurrences were in November—on the 1st and 7th in 1905; and on the 10th and 28th in 1903. The fifth was obtained on 27th June 1905—a remarkable date for its appearance.

Wheatear, *Saxicola œnanthe.*—Is a summer visitor to the group, and about four pairs nest on Eilean Mor. But the bird is chiefly a bird of passage, and as such is one of the commonest visitors to the islands.

The spring passages are recorded from 6th April to 29th May, but the birds are most abundant from mid-April to mid-May.

The return autumn movements do not commence in earnest until the second week in August, are at their height in September, and the last of the rearguard has been observed as late as 20th November (see also p. 257).

Whinchat, *Pratincola rubetra.*— There are two known visits only. A female occurred on 26th May 1906; and another on 25th September 1908. Both of these were sent for identification.

Stonechat, *Pratincola rubicola.*—We know of one occurrence only of this species—namely, a male on 3rd May 1906, which was duly forwarded.

Wren, *Troglodytes troglodytes.*—A somewhat frequent visitor, but never more than three have been seen on any occasion. It is impossible to say what the precise nature of these visits is, but they suggest that the bird may possibly be resident on one or other of the islands, for it has been seen in January, February, March, April, October, and November. We did not see it, but it might easily have escaped our notice on the face of steep cliffs, especially where the marguerites were abundant.

SPOTTED FLYCATCHER, *Muscicapa grisola.*—Single birds have been captured, and forwarded, on three occasions—viz., on 4th June 1906, on 14th June 1905, and on 22nd September 1909.

PIED FLYCATCHER, *Muscicapa atricapilla.* — See observations on p. 257. The further records are for two birds killed at the lantern (and sent) on 22nd September 1908.

SWALLOW, *Hirundo rustica.*—A few appear annually—chiefly in May, but occasionally in June, the dates ranging from 13th May to 3rd June.

There are only three records for autumn—namely, for 6th, 10th, and 13th September.

HOUSE-MARTIN, *Chelidon urbica.*—As in the case of the Swallow, a few visit the islands annually, mainly in the spring. They have been observed from 10th April to 18th June. On the latter date, one which had been killed by a Merlin was sent to me.

There are two records for autumn—4th August 1907 and 3rd November 1907.

SAND-MARTIN, *Cotile riparia.* — There are records for the occurrence of single birds on 23rd and 30th June 1901, but none are known to have appeared since.

SWIFT, *Cypselus apus.*—The Swift appears annually, late in spring, when single birds have occurred from

21st May to 28th June. On 15th June 1907, however, twenty appeared—a most unusual occurrence.

For autumn there are records of single birds at dates ranging from 16th to 28th September.

CUCKOO, *Cuculus canorus.*—Has been known to visit Eilean Mor twice in June. On these occasions, single birds were killed at the lantern on the 9th of the month in 1906, and on the 15th in 1907.

SHORT-EARED OWL, *Asio accipitrinus.*—Single birds have been recorded on five occasions during spring and autumn—namely, on 5th and 8th May, 5th June, 29th September, and 3rd November.

SNOWY OWL, *Nyctea nyctea.* — Single birds have been recorded on two occasions—namely, on 6th and 7th May 1906, and 12th October 1903.

PEREGRINE FALCON, *Falco peregrinus.*—In addition to the native pair of birds, which leave in September, the islands are occasionally visited by falcons of this species in the late autumn and winter.

MERLIN, *Falco æsalon.*—A regular visitor during the spring and autumn passage movements, probably accompanying its migratory prey on their journeys. It is also an occasional winter visitor.

Its spring appearances are dated from 2nd March, during April and May, and as late as 7th June.

In autumn it has been observed between 2nd September and 16th November. One was captured at the lantern on 19th September 1900, and sent; and another visited the light and was seen taking Wheatears flying in the rays on 2nd September 1905. On 30th September 1902, a Merlin chased a Pipit into the engine house and was secured.

GREENLAND FALCON, *Falco candicans.*—Has several

times visited the islands. An adult male appeared at Eilean Mor on 8th March 1908, and raided the Guillemots. The other records are: April 1907, one shot and remains sent; 18th April 1908, one; 2nd December 1909, one; another on the 14th; one on 25th January 1910; and lastly, one on 17th March 1910.

ICELAND FALCON, *Falco islandus.*—A female was shot about 10th December 1908, and sent in the flesh to Mr Fred Smalley of Challon Hall, Silverdale.

KESTREL, *Falco tinnunculus.*—An occasional visitor, whose appearances are chronicled in eight instances in the records. These refer to its occurrence in March, June, September, and November. A male captured on 18th June 1904 was forwarded.

CORMORANT, *Phalacrocorax carbo.*—Mr Harvie Brown saw several pairs on Bronna cleit on 11th June 1881—the only record.

SHAG, *Phalacrocorax graculus.*—Very common all the year round (see also pp. 258-9).

GANNET, *Sula bassana.*—Seen at all times during the year, but its visits are rare in winter. During the latter half of March, flocks and pairs are recorded as constantly passing the island, and proceeding in the direction of St Kilda.

HERON, *Ardea cinerea.*—Frequently seen in the autumn, when single birds, or never more than two, are observed. Has appeared from 9th August to 27th October. These visitors do not remain long, indeed they have a poor time of it, for the opportunities of obtaining food are *nil.*

"GREY GEESE," *Anser sp.* (?).—Grey geese are occasional visitors during the autumn and winter, but their identity has not yet been established.

BRENT GOOSE, *Branta bernicla.*—The single bird seen by us on 11th September 1904, is the only known instance of the occurrence of this species.

BERNACLE GOOSE, *Branta leucopsis.*—About two thousand are said to have their winter quarters on the various islands of the group. The date of their first appearance in autumn varies from 6th to 15th October, and they continue to arrive down to 10th November.

They are observed gathering for departure from 16th March, and the latest date for them in spring is that of a flock seen on 15th May.

TEAL, *Nettion crecca.*—There are a few records for the occurrence of single birds for January, April, July, and October.

MERGANSER, *Mergus serrator.*—The only recorded visit refers to a female which appeared on 20th April 1903.

EIDER DUCK, *Somateria mollissima.*—Breeds commonly, but is not present in winter, though a single bird was seen on 23rd January 1906.

The earliest record of its appearance in spring is 6th February, and it continues to arrive down to 6th April. The latest date on which it has been observed in autumn is 12th November.

RING-DOVE, *Columba palumbus.*—There are two records for the visits of single birds; one was found dead on Eilean Mor on 14th May 1906, and another was observed resting on 1st July 1904.

TURTLE DOVE, *Turtur turtur.*—There are six records of the visits of single birds during the autumn, the dates ranging from 11th September to 3rd October. On the first-named date the bird was obtained, and sent for determination.

CORN-CRAKE, *Crex crex.*—Is occasionally observed in spring and autumn. As there are several records for June, it is not impossible that the bird breeds on some of the islands. The dates for its earliest and latest appearances are 8th May and 16th September respectively.

WATERHEN, *Gallinula chloropus.*—One was captured on Eilean Mor on 1st November 1905.

PRATINCOLE, *Glareola pratincola.*—An adult female, now in the Royal Scottish Museum, was obtained on Eilean Mor on 13th July 1908. The visit of this Southern European summer bird affords another remarkable instance of the appearance of migratory birds at places far removed from their accustomed seasonal haunts, and much astray from the routes that should have been followed to reach them.

[GREY PLOVER, *Squatarola helvetica.* — There are several records of "Grey Plovers" for both spring and autumn, but as no example has been obtained, the visits of this species must for the present be regarded as unconfirmed.]

RINGED PLOVER, *Ægialitis hiaticola.* — Occurs at intervals, in small numbers, from mid-July to near the end of October. It occasionally visits Eilean Mor late in May and during June, indicating that it possibly nests on some of the adjacent islands of the group.

DOTTEREL, *Eudromias morinellus.*—It is recorded that a single bird visited the island on 12th June 1903, and that nine appeared and rested on 30th August of the same year. A bird of the year was killed in September 1906, and was sent for identification—the first authentic record for occurrence of this species in the Outer Hebrides—and an adult was captured on 31st May 1910.

GOLDEN PLOVER, *Charadrius pluvialis.*—A regular visitor on the spring and autumn passages in small numbers.

Has appeared in spring from 9th April to 19th June, being most numerous during the latter half of April.

It has been observed after the breeding-season as early as 29th July, but the autumn movements are chiefly witnessed in September and the first half of October; the latest date for their observation is 9th November.

LAPWING, *Vanellus vanellus.*—Occurs regularly on both passages, chiefly in small parties, rarely in flocks; and appears in winter during cold and snow.

In spring it is recorded from 12th March to 5th June, mainly from mid-March and during April; less frequently in May.

In autumn it has appeared between 13th September and 20th November, mostly in the latter half of September and during October.

The winter visits are nearly all chronicled for December.

TURNSTONE, *Strepsilas interpres.*—Appears regularly in autumn from 13th August to 19th November, and is observed occasionally in winter.

OYSTER-CATCHER, *Hæmatopus ostralegus.*—A common summer visitor to the islands for nesting; arrives in March, the 4th being the earliest record, and departs in August.

GREY PHALAROPE, *Phalaropus fulicarius.*—There are two records of visits paid in May 1906—a single bird on the 18th, and a pair on the 19th.

Woodcock, *Scolopax rusticula.*—Occurs regularly on the autumn passage, in small numbers, at dates ranging from 16th October to 18th November. It has been known to appear in December during snow.

There are no records of visits during the spring migration northwards.

Common Snipe, *Gallinago gallinago.*—A regular and frequent visitor in small numbers during the periods of passage.

In spring they pass on their way northwards from mid-March, during April, and sometimes as late as 11th May.

The autumn return movements commence in mid-September, and are in progress until late November. There are a few records of appearances in late July.

During cold snaps and snow it appears in considerable numbers, and in company with other species similarly affected; such visits have been known to occur as late as 27th February in 1905, when many appeared during snow.

Jack Snipe, *Gallinago gallinula.*—Though common and regular on its autumn passage, it has very seldom been detected on its way north in spring. Occasionally occurs in some numbers during cold periods in winter.

The autumn movements are chronicled as from 14th September to 20th November, the chief period being the latter half of September and during October. A remarkable visitation on 17th September 1904 has already been described (pp. 260-1).

The only spring record is for a single bird on 25th May 1907.

Dunlin, *Tringa alpina.*—A few appear annually in spring and autumn, and rarely in winter. It has occurred in autumn from 7th August to 12th October, and in spring from 15th May to 5th June.

On 9th September 1908, a male of the small Continental race, the *Pelidna schinzii* of Brehm, was obtained and forwarded for identification.

Purple Sandpiper, *Tringa striata.*—A winter visitor to the group. It has been known to arrive as early as 28th August, and to remain as late as 28th May.

Knot, *Tringa canutus.*—One killed at the lantern on 18th November 1902, and sent for determination, is the only known instance of a visit of the Knot to the islands.

Sanderling, *Calidris arenaria.*—There are records of the visits of single birds on two occasions—2nd June 1901 and 5th September 1905, when the bird was sent for identification.

Redshank, *Totanus calidris.*—The islands being devoid of marshy places, afford no attractions for this familiar wader, and hence the records of its visits are few.

Single birds are known to have appeared twice in spring, in April and May, and on three occasions in autumn, in August and September.

Curlew, *Numenius arquata.*—A regular visitor to the islands in small numbers during spring and autumn, and in cold periods in winter.

The spring passages date from early April to 20th May; and the autumn movements from 4th August to 18th November.

Whimbrel, *Numenius phæopus.*—The fact of its having bred has already been alluded to (p. 262). A few visit Eilean Mor, off and on, all through the summer, and the bird doubtless nests on some of the adjacent islands. As a bird of passage, it appears in flocks from 1st May to 5th June, returns, on its way south, during September, and has been seen as late as 4th October.

ARCTIC TERN, *Sterna macrura.*—A few are recorded as being off the islands in June.

BLACK-HEADED GULL, *Larus ridibundus.* — This species is of rare occurrence so far to the west, and at such pelagic haunts. There is only one record of its visit—namely, for 8th May 1906, when two were observed.

GREAT BLACK-BACKED GULL, *Larus marinus.*—Two pairs remain all the year on Eilean Mor. There are records of others appearing early in February, and it doubtless breeds on all the islands of the group affording suitable haunts. For other observations, see p. 262.

LESSER BLACK-BACKED GULL, *Larus fuscus.*—Some nest on Eilean Tigh. The date of their arrival in spring has not been noted, and they had departed previous to 6th September in 1904. For other observations, see p. 262.

HERRING GULL, *Larus argentatus.*—Nests abundantly on the uninhabited islands of the group. On Eilean Mor a few only nest, but a score remain all winter, attracted by the refuse thrown out from the lighthouse.

KITTIWAKE, *Rissa tridactyla.*—The commonest of the gulls during the nesting season. It arrives late in February, and departs before September. On 3rd November 1907, hundreds appeared, simultaneously with a rush of Redwings, Fieldfares, Blackbirds, Skylarks, and Snow Buntings; also with a few Woodcock, and a Short-eared Owl.

ARCTIC SKUA, *Stercorarius crepidatus.*—Two seen on 28th August 1903, the only record.

RAZORBILL, *Alca torda.*—Vast numbers visit the islands as breeding-stations, arriving according to the records, in some seasons late in February, and in others late in March; all had departed before we arrived on 6th September 1904.

COMMON GUILLEMOT, *Uria troile.*—Many thousands make their homes on the islands during the summer. They are said to arrive from the middle of February to the end of that month. They had quitted their summer haunts before I arrived on 6th September 1904.

BLACK GUILLEMOT, *Cepphus grylle.*—See remarks on this species at p. 263.

LITTLE AUK, *Alle alle.*—There are two records only for the visits of this species. In 1903 one was seen on 13th December, and in 1909 one was found on 12th November.

PUFFIN, *Fratercula arctica.*—Described as being in millions on the islands in summer. Arrives during the latter half of April, and had left, with a few exceptions, before our arrival on 6th September. See remarks on this species on p. 263.

STORM PETREL, *Procellaria pelagica.*—First noted in the spring on 4th May, and latest in autumn for 18th October. Our observations on this species will be found on p. 264.

FORK-TAILED PETREL, *Oceanodroma leucorrhoa.*—This species has already been noticed, and our experiences related on pages 264-5. The earliest notice in the schedules for spring is for 6th April. The egg has been found as early as 29th May. The light-keepers aver that it captures moths at the lantern.

FULMAR, *Fulmarus glacialis.*—Two pairs bred on Eilean Mor in 1904—the first instance of their breeding on that island. It is known, however, to have bred on some of the outer islets of the group since 1902. The earliest date for its appearance in the spring is 4th February, and the latest for autumn is 7th September.

MANX SHEARWATER, *Puffinus anglorum.*—Great numbers are scheduled as visiting the neighbourhood of the islands on 16th March 1901—the only recorded occasion.

GREAT SHEARWATER, *Puffinus gravis.*—The birds seen by us on the islands on 21st September 1904, are the only ones that have come under observation.

SLAVONIAN GREBE, *Podicipes auritus.*—The immature bird, already mentioned as observed on 16th September 1904, is the only record of the occurrence of this species.

CHAPTER XXV

SULE SKERRY, WEST OF ORKNEY, AND ITS BIRD-VISITORS

On leaving the Flannan Islands, I was much interested to learn that the lighthouse steamer *Polestar* was to proceed to Sule Skerry with stores ere she returned to Stromness. This was excellent news, for it would afford me an opportunity of visiting and exploring another outlying station, and of making the personal acquaintance of light-keepers who had for some years sent me records of the migratory birds visiting this remote islet. I landed on the island on the morning of 22nd September 1904, and spent half a day reconnoitring, searching for bird-visitors, and interviewing the keepers.

Sule Skerry is a small island lying out in the Atlantic, being 35 miles north-west of Hoy Head, Orkney, and 22 miles north-east of Cape Wrath, the north-western extremity of the mainland of Great Britain. It, no doubt, derives its name from its contiguity to the stack lying some 4 miles west-by-south, which is a great haunt of the Gannet, the Gaelic name for which is Sulaire. As the terminal portion of its name implies, Sule Skerry lies very low in the water, the highest portion, the centre of the island, being only 45 feet above high water, while

PLATE XXIV.

[*Photo: C. Dick Peddie.*

SULE SKERRY FROM THE SOUTH-EAST.

the broad belt of rugged rocks forming the protecting barrier or coast-line averages 15 feet, rising in one or two places to 30 feet. Thus it is not surprising that less than one-half its area of 35 acres is free from the onsets of the great Atlantic breakers during the prevalence of storms, which are of frequent occurrence in its latitude. This oasis forms the central and highest portion of the island, and is clothed in summer with a luxuriant crop of coarse grass, marguerites (*Chrysanthemum inodorum*), and other herbage. The marguerites flourish exceedingly among rough stones and rocks, most of which they completely hide, as he who ventures among them will soon find to his great discomfort.

As Sule Skerry was a considerable source of danger to navigation, and had been the scene of many disastrous shipwrecks, a lighthouse was erected on it, so that since the autumn of 1895 it has ceased to be a source of anxiety to mariners.

The island is visited annually by a number of migratory birds, probably those which traverse the west coast of Scotland and the Hebrides, when proceeding to and from Northern Europe *via* the Shetland Isles. Of these, a few alight to rest or to search for food. The great majority pass in the night, when, unless the atmospheric conditions at the moment are favourable for bringing the decoying powers of the lantern into display, nothing is observed. The fact that so many species, and so many individuals of some of them, have come under notice, goes far to establish the importance of the station. But while the island lies in the regular course of certain migrants, yet there are others which we should expect to find, such as the Starling and the Skylark, which are remarkable exceptions. The number of species which are *known* to have occurred (including native birds), has reached the extraordinary total of 103, and includes such rarities as the Northern Willow-Warbler, the Siberian Chiffchaff, and others.

During my short visit, I observed thirteen species of birds of passage, including two, the Lapland Bunting and the Grey Plover, which had not previously been known to occur.

The lighthouse observations date from 1899 to the present time, and are the source of practically all our knowledge of the Sule Skerry bird-visitors, which, thanks to the valuable co-operation of the light-keepers, is considerable. In this connection the contributions of the late Mr James Tomison, a most painstaking and capable observer, call for special acknowledgment, and form the mainstay of this chapter. Mr Tomison also published in the *Annals of Scottish Natural History for* 1904, an excellent paper on the birds of this island, a contribution to which I shall have frequent occasion to refer.

CORVUS CORNIX, *Hooded Crow.*—A few appear at intervals during the spring and autumn. Many, however, visited the island on 2nd March 1901, from which date to 11th April the bird has appeared in the spring, sometimes in company with Rooks. The autumn records, which are few, bear dates from 22nd October to 4th November.

CORVUS FRUGILEGUS, *Rook.*—Is almost annually recorded as a visitor in spring, at dates ranging from 26th February to 28th April; the birds sometimes appear in large numbers, and in company with Hooded Crows. It has only once been observed in the autumn.

CORVUS MONEDULA, *Jackdaw.*—There are no data

relating to the appearance of this bird in the schedules, but Mr Tomison mentions that Jackdaws, along with Rooks and Grey Crows, are sometimes driven out of their course by south-east winds in spring.

STURNUS VULGARIS, *Starling.*—There is no evidence that Starlings on passage visit the island, which is a somewhat remarkable fact. A few appear irregularly during most months of the year, twelve being the largest number observed on any occasion.

FRINGILLA CŒLEBS, *Chaffinch.*—This bird is not a regular visitor during the passage seasons. A few have appeared late in March, during April, and early May, and again in October and November.

FRINGILLA MONTIFRINGILLA, *Brambling.*—This species is quite an uncommon visitor. It has been known to appear on three occasions in spring, at dates ranging from 29th April to 29th May. On the night of 5th October 1906, several appeared at the lantern, and next morning forty were observed on the island—the only records for autumn.

CHLORIS CHLORIS, *Greenfinch.*—There is one record only, namely, the visit of twelve birds on 4th November 1906.

ACANTHIS FLAVIROSTRIS, *Twite.*—There are no data in the schedules, but Mr Tomison says that it is most commonly seen in April and May, sometimes in large flocks, which remain for a few days. Small numbers appear in August and September and occasionally in November.

ACANTHIS LINARIA, *Mealy Redpoll.* — Redpolls are recorded as having visited the Skerry on several occasions in autumn, and at dates ranging from 15th September to 5th November. In some instances examples were sent, and belonged to the typical form.

ACANTHIS ROSTRATA (ACANTHIS LINARIA ROSTRATA), *Greater Redpoll.*—One was captured at the lantern during the second week of October 1911, and was sent for determination.

PASSER DOMESTICUS, *House Sparrow.*—Mr Tomison tells us that this bird is seen at intervals for a few days about midsummer—a remarkable circumstance, but vouched for by a most accurate and capable observer.

LOXIA CURVIROSTRA, *Crossbill.* — Sule Skerry came within the area invaded by the remarkable irruption of Crossbills which swept over our islands during the summer of 1909. Their incursion into this remote islet was heralded by the appearance and capture of a bird at the lantern on 28th June. More followed, and forty-two was the largest number observed. They remained on this bleak, and indeed desert island, from a Crossbill's point of view, for about three weeks. Mr Moore, the light-keeper, tells me that he found several dead amid the territory occupied by the Arctic Terns, and he is of opinion that the Terns put them to death for venturing within the precincts of their domain.

EMBERIZA MILIARIA, *Corn - Bunting.* — Appears irregularly. A large flock arrived in December 1899, and remained nearly a month.

EMBERIZA PUSILLA, *Little Bunting.*—One was killed at the lantern on 22nd September 1908, and sent for identification—the second Orcadian record.

CALCARIUS LAPPONICUS, *Lapland Bunting.*—I saw several of these birds on the island during my visit on 22nd September 1904.

PLECTROPHENAX NIVALIS, *Snow-Bunting.*—Is a common bird of double passage, and a winter resident in small numbers. The spring passage commences at mid-March, and lasts until 2nd June. The chief movements are chronicled for the last week of March. The autumn visits date from 11th September to 14th November, large flocks appearing in October and early November.

ALAUDA ARVENSIS, *Skylark.*—Visits the islands in spring, in some seasons in fair numbers, from 2nd March to 8th April. Strange to say, there are no records whatever of its occurrence in autumn; but the bird has several times appeared in January. Mr Tomison says that it is rarely absent in spring and summer, but does not nest.

MOTACILLA LUGUBRIS, *Pied Wagtail.*—This species has been much confused in the records with the White Wagtail, and I am only able to say that there is a single undoubted occurrence of the pied species at Sule Skerry—namely, one on 30th August 1906.

MOTACILLA ALBA, *White Wagtail.*—A regular visitor on both the spring and autumn passages. Its appearances date from 28th April to 8th May in spring; and between 23rd August and 24th September in autumn. I saw several on the island on 22nd September 1904. It is not recorded as being numerous.

ANTHUS PRATENSIS, *Meadow-Pipit.*—Occurs on passage in April, August, and September. The records are, however, not numerous, and refer to visits of small parties. I saw several on the island on 22nd September 1904.

ANTHUS OBSCURUS, *Rock-Pipit.*—Resident all the year round, but not numerous. They must, says Mr

Tomison, emigrate, as their numbers do not seem to increase. I have no doubt the Merlin exacts a heavy toll from this species.

Regulus regulus, *Goldcrest.* — As at all far-western stations, so at Sule Skerry, this well-known little bird is a very uncommon visitor. During eleven years it has only been recorded as having appeared on four occasions, all in autumn. These visits date from 29th September to 15th October, and relate to few individuals, fifteen being the largest number seen.

Turdus musicus, *Song-Thrush.*—Occurs occasionally on passage in autumn, sometimes in rushes, but has very seldom been seen in spring. The data for autumn, which are slight, indicate that the movements are performed from 25th September to 24th October.

Turdus iliacus, *Redwing.*—A visitor during its passages, which are recorded for spring from 31st March to 22nd May, and in autumn for 20th September to 14th November. In the former season these visits are most frequent, and the migrants most numerous, during April; and at the latter great rushes are witnessed from mid-October to early November.

Turdus pilaris, *Fieldfare.*—Is a visitor during the autumn passage, occurring in October and November, sometimes in great rushes. The earliest date for its appearance is 5th October in 1905. In spring it is seldom noticed, the dates of the few records ranging from 9th April to 6th May. It has appeared on a few occasions in December and January, during snow and severe weather.

Turdus merula, *Blackbird.*—Appears, sometimes in great numbers, in autumn, at dates from 8th October to 14th November. It is less numerous on the spring passage, when its visits have been from 24th March to 24th May. There are also records of appearances on 28th December, during snow, and on 24th February.

TURDUS TORQUATUS, *Ring-Ouzel.*—Occurs on both the spring and autumn migrations, but is not recorded for all years. It has passed north in spring from 12th April to 16th May, and on its return passage from 9th September to 16th October, when it has sometimes occurred in considerable numbers.

ERITHACUS RUBECULA, *Redbreast.*—A few appear in autumn, but the records are scanty, and relate to visits paid from 13th September to 28th October.

SAXICOLA ŒNANTHE, *Wheatear.* — Regular and abundant as a visitor during passage. Appears in spring from 10th April to 11th May; and in autumn from 6th August to 6th October. Some of the numerous visitors, no doubt, belong to the large north-western race, *Saxicola œnanthe leucorrhoa.*

RUTICILLA PHŒNICURUS, *Redstart.*—There are only three autumn records for the occurrence of this species—namely, for 20th and 22nd September and 6th October. On the latter date two were captured at the lantern, and sent. The only spring visit noted was that of two birds on 15th May 1910.

SYLVIA CURRUCA, *Lesser Whitethroat.*—One was killed at the lantern on the night of 17th September 1902, and forwarded to me in the flesh. This species has been detected as a visitor on migration at all the Scottish island stations—even on this remote rocky islet—and yet it is very seldom detected on passage on the mainland of Scotland.

SYLVIA BORIN, *Garden-Warbler.*—There are three occurrences, all for September, from the 5th to the 25th.

One obtained at the lantern on the last-named date was forwarded for determination.

PHYLLOSCOPUS SIBILATRIX, *Wood-Warbler.* — This species has on few occasions been observed northwards of the mainland of Britain. The first of these was a bird killed at the Sule Skerry lantern on the night of 27th September 1906, and sent to me for determination.

PHYLLOSCOPUS TROCHILUS, *Willow-Warbler.*—Judging from the records, this must be a rare visitor, for there are only two instances of its appearance registered, both for September, and at the lantern. The birds were sent. Mr Tomison makes no allusion to the occurrence of this species in his paper on the avifauna of the island.

PHYLLOSCOPUS BOREALIS, *Northern Willow-Warbler.* —An adult male was captured at the lantern on the night of 5th September 1902, and was forwarded to me in the flesh. At the time, this bird was believed to be an example of the Greenish Willow-Warbler (*P. viridanus*), and was exhibited as such at a meeting of the British Ornithologists' Club. Later, in September 1908, I shot a specimen of *P. borealis* at Fair Isle (see p. 130), and this led to the detection of the error in the identification of the Sule Skerry bird. The Sule Skerry and Fair Isle birds are the only ones known to have occurred in the British Isles.

PHYLLOSCOPUS COLLYBITA, *Chiffchaff.* — One was killed at the lantern on the night of 3rd November 1901, and was sent for identification.

PHYLLOSCOPUS TRISTIS, *Siberian Chiffchaff.* — The first British example of this North-Eastern European and Siberian summer bird, was killed at the Sule Skerry lantern on the night of 23rd September 1902.

Luckily it was placed in spirit, and sent to me in the flesh, by the next relief steamer visiting the island, otherwise the species would have escaped detection within the British area for some years. The whole of the circumstances relating to the occurrence of this inconspicuous little bird on that remote Atlantic rock, where, by the merest chance, it was saved from oblivion, forcibly remind us how very many similar cases there must be annually of migrants escaping notice, including the great majority of the commonest species—a fact that unfortunately many interested in the subject of migration do not fully realise, and hence the rash speculations which are raised on the very flimsiest of foundations.

TROGLODYTES TROGLODYTES, *Common Wren.*—One appearance on the island, in September 1900, is the only known visit.

MUSCICAPA ATRICAPILLA, *Pied Flycatcher.*—Single birds have occurred on three occasions—twice in September and once on 5th November 1901. The latter is a very late date for the appearance of this species anywhere in the British Islands.

HIRUNDO RUSTICA, *Swallow.* — According to Mr Tomison, a few are seen every year in May and June.

CHELIDON URBICA, *House-Martin.*—A casual visitor whose appearance has been recorded on four occasions. These are 16th May, 15th June, 22nd June, and 27th August, and four is the largest number observed.

COTILE RIPARIA, *Sand-Martin.*—One is recorded for 21st May 1910.

CYPSELUS APUS, *Swift.*—There are nine recorded appearances of this species, eight of which were in August, at dates ranging from 12th to 30th ; and one on

4th July 1901, when eleven, the largest number observed, appeared.

CUCULUS CANORUS, *Cuckoo.*—Is a very uncommon visitor, one which has only been recorded on five occasions in eleven years. These records were made in the spring, at dates ranging from 8th May to 8th June, and relate to the appearance of single birds.

DENDROCOPUS MAJOR, *Great Spotted Woodpecker.*—Two appeared on the island on 24th September 1905, the only record.

ASIO ACCIPITRINUS, *Short-eared Owl.*—There are autumn records of three occurrences late in October. These relate to the visits of one or at most two birds. The only bird seen in spring appeared on 30th April 1910.

FALCO TINNUNCULUS, *Kestrel.*—Evidently a rare visitor, there being only three entries of its occurrence in the records. These refer to the appearance of single birds in February, August, and September.

FALCO ÆSALON, *Merlin.*—A frequent visitor during the period when small birds are abundant on passage in the autumn—*i.e.*, from late in August until November. It seldom appears in spring, but one is recorded for 5th April 1907, and another for 22nd May 1910.

ARDEA CINEREA, *Heron.*—This is an irregular and infrequent visitor, but has been observed during most months of the year, more especially in August, September, and October. As a rule, single birds are seen, but on 15th October 1899, five appeared.

PHALACROCORAX GRACULUS, *Shag.*—An abundant resident (Tomison).

ANSER ALBIFRONS, *White-fronted Goose.*—One was observed feeding on the island on 3rd May 1910.

BRANTA LEUCOPSIS, *Bernacle-Goose.* — Has been occasionally observed moving north at the end of April, and southwards after the middle of October.

TADORNA CASARCA, *Ruddy Sheld-Duck.*—An adult female was obtained on 18th June 1909, and was forwarded to me for identification. This is the first known instance of a visit of this bird to any of the northern isles.

ANAS BOSCAS, *Mallard.*—Is an occasional visitor, usually in pairs, during the winter months (Tomison).

NETTION CRECCA, *Teal.*—Far from a common visitor, according to Mr Tomison and the records. On 27th February 1906, a male was shot and sent to me; two appeared on 30th March 1911; and one was observed on 21st October 1908.

MARECA PENELOPE, *Wigeon.*—A few visit the island every year, according to Mr Tomison, but there are no entries relating to this bird's appearance in the schedules.

FULIGULA FULIGULA, *Tufted Duck.*—There is one known occurrence only—that of a male on 30th April 1902, whose wings were sent me for identification.

FULIGULA FERINA, *Pochard.*—Two appeared on 13th March 1903, and the wings of one of them were sent for determination.

FULIGULA MARILA, *Scaup.*—There are three entries in the schedules for the visits of this species. These relate to the appearance of single birds in February, August, and October.

HARELDA GLACIALIS, *Long-tailed Duck.*—A young male, shot on 19th October 1901, is the only known appearance of this common winter visitor to our northern seas.

SOMATERIA MOLLISSIMA, *Eider Duck.*—Resident, but

only a few are seen in November and December. Numbers appear early in January, and spend all their time afloat, no matter how stormy the weather may be (Tomison).

MERGUS SERRATOR, *Merganser.*—Mr Tomison only once saw this bird, and there have been no other instances recorded since he left the island.

CREX CREX, *Corn-Crake.*—Occurs regularly in May and occasionally in September. In spring the earliest date for its appearance is 3rd May, and the last for the autumn 25th September. According to Mr Tomison, it is seen in June and July, but does not nest, and rarely "crakes."

RALLUS AQUATICUS, *Water-Rail.*—Mr Tomison several times observed this bird in winter. One is recorded as having been seen on 20th June 1909.

GALLINULA CHLOROPUS, *Waterhen.*—One was killed at the lantern on 12th April 1910.

CHARADRIUS PLUVIALIS, *Golden Plover.*—Occurs on passage in both spring and autumn. The spring records for its flights northwards date from 2nd April to 3rd May; and the return autumn movements have been witnessed from 20th August to 30th November. Large flocks are heard passing at night during the latter season, and many fly around the lantern, but Mr Tomison has only known one bird to strike. Few are seen on the island during the daytime.

SQUATAROLA HELVETICA, *Grey Plover.*—I saw a single bird on the island on 22nd September 1904—the only record.

VANELLUS VANELLUS, *Lapwing.*—Occurs regularly, but in small numbers only, during the spring and autumn passage movements. In spring it has appeared from the end of February until 16th May; but March is the main month for its visits at this season. As a straggler it has been seen in June and July. The return migration dates from 20th August to 4th November. There are one or two instances of visits of a few birds in January and early February.

ÆGIALITIS HIATICOLA, *Ringed Plover.*—There is only a single record for the visit of this species; but that it should be so extremely rare is remarkable. Four were seen on 18th August 1906.

HÆMATOPUS OSTRALEGUS, *Oyster-catcher.*—The skerry, with its broad belt of low rocks, affords an ideal nursery for this species, and hence it is a summer resident in considerable numbers. The harbingers of the nesting season appear in the second week of February, and continue to arrive until the end of April. All leave early in the autumn, and I did not see a single Oyster-catcher during my visit on 22nd September 1904.

STREPSILAS INTERPRES, *Turnstone.*—A winter resident and bird of passage. The winter visitors have been known to arrive as early as 20th July, and depart in April. Mr Tomison relates that these winter birds are very tame, and come to the vicinity of the lighthouse to be fed.

The visits on passage in spring date from 6th May to 4th June, during which period considerable numbers are observed on their way north. In autumn the main movements are observed in August and September.

PHALAROPUS HYPERBOREUS, *Red-necked Phalarope.*—A single record only: one was captured at the lantern on the night of 29th October 1908, and sent for identification.

PHALAROPUS FULICARIUS, *Grey Phalarope.*—Has

occurred three times during winter. One was killed at the lantern (and sent) on 17th December 1908; another was observed off the island on 5th January 1909; and a female was captured (and sent) on 15th February 1904.

SCOLOPAX RUSTICULA, *Woodcock.*—A regular visitor on its autumnal migrations, sometimes appearing in considerable numbers. The dates of these visits range from 30th September to 10th November. A few have occurred during snow, in December, January, and February.

It is irregular in its visits when migrating north in spring, but has been observed from 28th March to 11th May.

GALLINAGO GALLINAGO, *Common Snipe.*—A winter resident, arriving late in August and during September, and departing in April.

Also a visitor on its spring and autumn passages, appearing during April (but has occurred as late as 11th May), September, and October.

GALLINAGO GALLINULA, *Jack Snipe.*—I saw two on the island on 22nd September 1904, and these, along with one recorded for the 20th of the same month in 1906, are the only known visits of this species. It is probably not so infrequent in its visits as the records would lead one to suppose.

TRINGA ALPINA, *Dunlin.*—There are no entries in the schedules relating to this species, but Mr Tomison remarks that it comes periodically. I saw several during my visit on 22nd September 1904.

TRINGA MARITIMA, *Purple Sandpiper.*—A winter resident. The earliest visitors arrive in August from the 4th onwards; the bulk appear in September. They take their leave of the island in April.

Totanus calidris, *Redshank.*—A visitor during many months of the year, but most frequent before and after the nesting season; that is in April, August, and September. On 23rd July 1901, thirty appeared, the largest number recorded. I saw several on 22nd September 1904.

Totanus nebularius, *Greenshank.*—Has occurred occasionally in both spring and autumn. I have received specimens which were killed at the lantern on 10th April 1905, and on 19th September 1905.

Limosa lapponica, *Bar-tailed Godwit.*—The only known visit of this species occurred on 13th May 1902, when a single bird appeared.

Numenius arquata, *Curlew.*—A small number winter on the island. These cold weather residents are remarkably constant in observing the times of their comings and goings, for they are recorded as making their first appearance from 1st to 8th August, and as taking their departure from 25th to 28th April. No evidence of passage movements is contained in the schedules, but these, no doubt, occur.

Numenius phæopus, *Whimbrel.*—A few are observed on the spring and autumn passages. These occur at the former season from 26th April to 9th May, when parties of two or three and small flocks are observed on their way northwards. The return autumn movements are recorded as from 26th July to 11th September. Mr Tomison mentions that about a dozen non-breeding birds spend the summer on the island.

Sterna cantiaca, *Sandwich Tern.*—One which visited the island on 25th June 1910 was killed by the Arctic Terns.

Sterna macrura, *Arctic Tern.*—Nests in great

numbers, and next to the Puffin is the most abundant species. The earliest date for its arrival is 10th May, and the latest 17th May. It begins to leave late in August, and by the middle of September all have gone.

Birds from further north occur on passage late in September, always during the night.

LARUS RIDIBUNDUS, *Black-headed Gull.*—Is generally seen for a day or two in the middle of summer (Tomison). One appeared on 6th May 1910.

LARUS FUSCUS, *Lesser Black-backed Gull.*—Not a common visitor. It usually appears in May, and does not remain long (Tomison).

LARUS MARINUS, *Great Black-backed Gull.*—About a dozen pairs are resident (Tomison).

LARUS ARGENTATUS, *Herring Gull.*—Some twelve pairs nest and remain all the year round. Many visit the island during the herring season (Tomison).

LARUS GLAUCUS, *Glaucous Gull.*—Fairly common during November, December, and January. Immature birds arrive in November, adults in December (Tomison).

LARUS LEUCOPTERUS, *Iceland Gull.*—According to Mr Tomison, it appears in November, December, and January.

RISSA TRIDACTYLA, *Kittiwake.*—A few pairs nest, and many visit the island on passage before and after the breeding season. The breeding birds arrive at the end of March, and leave early in August. A great number appeared on 1st June 1902, on their way north; and large flocks are seen on their return in July and August.

STERCORARIUS PARASITICUS, *Buffon's Skua.*—An adult male was captured on 9th June 1908, and sent to me.

PROCELLARIA PELAGICA, *Storm-Petrel.* — Many nest

amid the rough stones; but Mr Tomison remarks that it is difficult to determine the dates of their arrival and departure. They have, however, been observed in large numbers on 1st June, and have been known to visit the lantern in November.

OCEANODROMA LEUCORRHOA, *Fork-tailed Petrel.*—Has appeared at the lantern on a number of occasions in May, August, and September.

FULMARUS GLACIALIS, *Fulmar.*—This species nests in some numbers at Hoy Head, and hence its occasional appearances off Sule Skerry from February to September are not surprising. On 11th February 1909, one was killed at the lantern during clear weather.

ALLE ALLE, *Little Auk.*—Mr Tomison tells us that flocks usually appear off the island in January, especially after a northerly gale. On 16th January 1898, two were found dead at the base of the lighthouse, having struck either the lantern or the tower.

ALCA TORDA, *Razorbill.*—A considerable number breed on the island. They rarely land before 1st May, and leave in August, sometimes as early as the 6th.

URIA TROILE, *Common Guillemot.*—Does not breed, and very few are observed as visitors off the island (Tomison).

URIA GRYLLE, *Black Guillemot.*—Mr Tomison records this species as a regular summer visitor, arriving off the island in February and March, but not landing for breeding until the end of April. Single birds are occasionally seen in November and December, but as a rule they do not winter in the vicinity of the island.

FRATERCULA ARCTICA, *Puffin.*—Mr Tomison tells us that enormous numbers make the island their home in summer. The average date of their first appearance is

10th April (7th earliest), and of their first landing 20th April (earliest 16th), for the birds spend from eight to twelve days at sea off the island before seeking *terra-firma.*

The first leave about the middle of July, and practically all have left ere the month of August has run its course.

PODICIPES AURITUS, *Slavonian Grebe.*—Two occurrences have been chronicled, of single birds, in October and December, and one of these was forwarded for determination.

PODICIPES FLUVIATILIS, *Little Grebe.*—One was shot on 4th November 1907, and sent for identification—a remarkable occurrence at so decidedly pelagic a station.

USHANT: THE SOUTH-WEST COAST AND PHARE DE CREACH.

[*To face p.* 304.

CHAPTER XXVI

(1) THE ISLE OF USHANT AS A STATION FOR OBSERVING BIRD-MIGRATION: BIRD-WATCHERS WATCHED! (2) ON MIGRATORY BIRDS OBSERVED AT ALDERNEY, CHANNEL ISLANDS

In Europe, Ushant or Ouessant occupies a position second to none as an observatory for witnessing the phenomenon of bird-migration.

It is an island lying right in the main stream of the feathered flood which annually rushes first north and then south, according to the season, along the coast of South-Western Europe, traversing one of the most important migration routes in the Old World.

Ushant, however, has a further interest, inasmuch as it is the junction where migrants in spring branch off on their various lines of flight to reach England, or to proceed along our Atlantic and Irish coasts, both shores of the Channel and those of the North Sea en route for their northern summer homes—their native lands. In the autumn it is the station where the same wayfarers and their young reach the western trunk line en route for their southern winter retreats (see map, Plate II.).

As nothing whatever appeared to be known concerning migration at this most promising island for the bird-watcher, I decided to make Ushant the scene of my autumn vacation in 1898; and I was fortunate enough to secure as a colleague my friend, Mr T. G. Laidlaw.

We arrived at the little port of Le Conquet, 12 miles west of Brest, the place of embarkation for Ushant, on the night of the 7th September 1898. Early the following morning we were on board the small steamer which conveys the mails to the Islands of Molène and Ushant—isles which are chiefly associated in the public mind with the loss of that ill-fated liner, the *Drummond Castle*, in 1896. The morning, however, was unpropitious, inasmuch as a dense sea-fog prevailed, and at mid-day the passage was abandoned until the morrow.

Fortunately at Le Conquet there was an estuary, formed of the embouchures of several small rivers. Here at low water there are extensive mud-flats, studded with several small islands which are then accessible from the shore. On these attractive feeding-grounds we spent several hours, observing Turnstones, Dunlins, Curlew-Sandpipers, Knots, Redshanks, Greenshanks, Common Sandpipers, Ruffs and Reeves, Whimbrels, and Curlews. In addition, a number of other species of migrants on passage, such as White Wagtails, Yellow Wagtails (a large flock), and Wheatears came under notice; and a Chiffchaff was heard in autumn song in a garden close to the shore.

On the morning of the 9th, we were under way at six o'clock, and were soon threading our course through those rock-studded and dangerous seas which lie between the mainland and our island goal. A glance at the chart of these waters shows that there extends from the mouth of the Gulf of Brest, in a north-westerly direction, a series of islands, innumerable islets, rocks, and reefs—many of the latter being just awash or partially submerged. This archipelago culminates in the comparatively large island of Ouessant, which we call Ushant. Through the numerous straits and channels, with which the group abounds, a tide rushes at the rate of fourteen knots an hour, rendering the surface of the sea for many miles a series of broad rapids and eddies, resembling the waters of a mighty river rather than those of the ocean. Add to this the remarkably rugged outlines of the stacks and many of the islets, and the quaintness that marks the little out-of-the-world community that dwells upon the island of Molène, and we have scenes which it would not be easy to match elsewhere in European waters.

The birds seen on the voyage were Manx Shearwaters, Cormorants, Shags, Herring-Gulls, Lesser and Greater Black-backed Gulls, Sandwich and Lesser Terns; and among migrants a Common Sandpiper and an Osprey.

A three-hours' run brought us to Ushant, the most westerly land of France, situated 12 miles from the nearest point of the mainland. This island is irregular in outline, for it throws out, as it were, two long parallel arms to the south-west, which enclose the deep Baie de Porspaul; while there are minor promontories to the north-east, north, and north-west. It is about $3\frac{1}{2}$ square miles in area, and does not present any remarkable physical features excepting the wonderful rock-scenery on the west coast, of which more anon. Cliffs face the sea on all sides except the south-west, and these attain their maximum height of 211 ft. in the north. The surface of the island has a parched and arid appearance, due to its herbage being closely cropped by the cattle and sheep of the inhabitants. There are, however, several shallow, verdant depressions, down which, no doubt, tiny streams may find their way to the sea during the winter rains. Some corn is grown, and whins flourish in compounds erected for their protection, for these shrubs form an important item of fuel. There are a few trees of small size in one or two of the gardens at Lampaul, the chief hamlet, which lies at the head of the Baie de Porspaul.

The west coast is exposed to the full force of the Atlantic, and by the fury of the waves the numerous more or less lofty rocks that stud the shore and the irregular face of its cliffs have been carved and transformed into all manner of remarkable and fantastic forms. Indeed, the rock-scenery on the west coast is wild and weird in the extreme.

Geologically, the island is composed of foliated granite. This rock in weathering does not form ledges, and this, in a measure, may account for the absence of breeding-stations of sea-fowl on the sheltered eastern cliffs; those on the north and west are storm-swept even in summer, and are consequently not available.

Just off Ushant lie many hundreds of islets and rocks, only a few of which are of considerable size, but none of them appear to be tenanted during the breeding-season by Gulls, Terns, or other marine species. On the large island of Balanec, which lies some 5 miles south-east of Ushant, many "Hirondelles de Mer," "Perroquets de Mer," "Goëlands," etc., rear their young—at least so we were informed.

The island has the surprisingly large population of over two thousand inhabitants. All the men are engaged in the lobster-fishery and appeared to be well-to-do; for a rich and practically inexhaustible harvest of these crustaceans lies among the vast series of submerged reefs and over the rocky sea-bottom, which extend for miles around Ushant.

There are two lighthouses on the island, namely, the Phare du Stif in the north-east, and the Phare de Creach in the south-west. The latter has long been lighted with electricity, and throws out powerful and rapidly revolving beams which can be seen for many miles.

At Ushant we had hoped to remain for several weeks, but we had only been six days on the island when an immigrant, radiant in blue and white, arrived; alas, not a feathered visitor, but a sergeant of gendarmes (there are no police on the island). This myrmidon of the law dogged our footsteps, at close quarters, during our rambles, while our place of abode was under his surveillance early and late. At first we were not disposed to take any notice of his presence, but the espionage at length became intolerable. And this was not all; the natives who had hitherto been most friendly, courteous, and obliging, not unnaturally regarded us with suspicion and avoided us, and our host at the little inn, although we had arranged to be his guests for some weeks, was wishful that we should depart. The situation had thus become insufferable, and we reported the matter by telegram to the British Consul at Brest, whose acquaintance we had made before leaving that city, and requested him to protest to the French authorities against the vexatious treatment to which we were being subjected. This the Consul most obligingly did, but his efforts were unavailing, for the authorities informed him that the gendarme had been sent to watch us by special instructions from Paris.

The Consul urged us to quit the island as soon as possible in order to avoid serious consequences. We left the island on 17th September, our *bête bleu* accompanying us, and returned to Brest. Thus was our Ushant expedition miserably wrecked.

It may be well to state here that our Foreign Office had informed the French Government of our intended visit to Ushant, and its object, some time before our arrival on the island. Thus the treatment meted out to us is inexplicable, except that our relations with France had just passed under a deep shadow, for the war cloud of Fashoda had appeared simultaneously with our advent on the island.

The following extract from the newspaper *La Patrie* throws some light upon the occurrence. Writing soon after our visit, this paper stated, on the authority of its Brest correspondent, that "the English are in the habit of visiting Ushant with a view to secure pilots well acquainted with these dangerous seas, and to bribe the islanders with British gold. Only last year, under the pretence of rewarding the islanders for their conduct in connection with the wreck of the *Drummond Castle*, they scattered a perfect golden shower over the islands. In short, our neighbours, in the time of peace, pave the way for the purchase of traitors in the time of war." This, we were credibly informed at Brest, was the true explanation. We were regarded as spies. Our bird-watching was a mere subterfuge. A party of engineer officers were engaged in surveying in connection with the fortifications it was intended to erect on the island, but the sites of these were not manifest in any way.

They professed much friendship, but it was to them, no doubt, that we owed our misfortunes.

During our short sojourn on the island we observed a number of birds. On these I shall now proceed to make a few remarks, reserving for the concluding annotated list the detailed particulars.

The following were doubtless resident species on Ushant:—Raven, Sparrow, Linnet, Corn-Bunting, Yellow Bunting, Skylark, Meadow-Pipit, Rock-Pipit, Stonechat, Redbreast, Hedge-Accentor, Wren, Peregrine Falcon, Ringed Plover, and, perhaps, the Oyster-catcher.

The summer visitors appeared to be only two in number—the Whitethroat and the Swallow.

The birds of passage observed between the 9th and 17th of September—a period of phenomenally fine weather, be it remarked—were the Redwing, Wheatear, Whinchat, White Wagtail, Grey Wagtail, Yellow Wagtail, Hen-Harrier, Sparrow-Hawk, Kestrel, Osprey, Heron, Turtle-Dove, Dotterel, Lapwing, Turnstone, Sanderling, Common Sandpiper, Redshank, Whimbrel, Curlew, Arctic, Common, Lesser, and Sandwich Terns, and Manx and Great Shearwaters. The absence of suitable shores on which to feed and rest accounted, no doubt, for the absence of several of the species of wading-birds which we had observed on the mud-flats of the opposite coast at Le Conquet. The brilliant weather, too, was decidedly against any migratory movement, pronounced or otherwise. Some of the species were, however, observed in considerable numbers.

The following species noted may perhaps be best described as winter visitors to the island, though the nesting-grounds of some of them are probably not far distant. These were the Kingfisher, Cormorant, Shag, Common Gull, Herring-Gull, Lesser and Greater Black-backed Gulls, and Kittiwake. Certain of the species observed as birds of passage—the Turnstone, for instance—would also be winter residents.

I took with me to Ushant a carefully prepared series of questions relating to the visits of migratory birds to the island, copies of which I left with the "Garde en Chef" of each lighthouse, with a request for answers and such other information bearing upon the subject as they could afford me. These documents I had to abandon on quitting the island; but on mentioning the fact to Consul Hoare, he most kindly offered to see the authorities at the Ponts et Chaussées, under whose jurisdiction the lighthouses fall, and endeavour to procure for me the information I desired. As the result, I received excellent and most useful answers to my enquiries from each of the Ushant lighthouses.

From these we learn that the island is visited annually by vast numbers of birds of passage; and that on dark, moonless and starless nights, with easterly winds, during the autumnal migratory period, from 500 to 600 birds are killed at the lantern—among others, Chaffinches, Thrushes, Blackbirds, Wild Ducks, Water-hens, Plovers, Lapwings, Woodcock, Snipe, and Curlews. As an illustration of the phenomenal numbers which sometimes occur, the Chef du Phare de Creach reported that on one night in the autumn of 1888, no fewer than 1500 birds perished by striking the lantern—an extraordinary number, but its accuracy is confirmed and vouched for by the engineer to the lighthouse authorities, to whom the Consul obligingly referred, at my request.

These reports also tell us that the greatest number of migrants appear in October, and it is then that *les grandes volées* occur. The following species are mentioned as occurring annually, excluding those already given as coming under the notice of Mr Laidlaw and myself:—Black and Grey Crows, Starlings, Chaffinches, Goldfinches, Siskins, Bullfinches, Buntings, Goldcrests, Warblers, Thrushes of various species, Martins, Cuckoos, Owls, Falcons, Herons, Wild Geese, Wild Duck, Teal, Ring-Doves, Quails, Land-Rails, Water-Rails, Waterhens, Golden Plover, Grey Plover, Woodcocks, Snipe, Sandpipers, and Gulls of various kind. In addition to these, M. Lucas, the Chef du Phare du Stif, very pertinently remarks that "L'île est encore visitée par d'autres espèces d'oiseaux, particulièrement de l'ordre des passereaux, mais dont les noms me sont inconnus."

Both of these observers agree that fewer migrants are observed in the spring, though the same species appear at that season.

In winters of great cold, and when the mainland is under snow, immense numbers of Starlings, Chaffinches, Thrushes, Blackbirds, Wild Geese, Wild Ducks, Teal, Quails, Water-Rails, Waterhens, Plovers, Lapwings, Snipe, and Woodcock, accompanied by Hawks of various species, seek the milder climate of the island, and usually remain until the end of February. In mild, moist, rainy winters, very few birds indeed visit the island.

According to the same authorities, the following birds nest annually on Ushant:—Ravens, Sparrows, Linnets, Larks, Pipits, Redbreasts, Warblers (whitethroats), Wrens, and Swallows.

How very much I wish I could have interviewed

these intelligent and well-informed observers. I did call upon them, but they courteously informed me that they could not afford me the information I desired, until they obtained permission to do so from headquarters.

There can be no doubt that had we been permitted to remain in peace upon the island until the early days of October, we should have obtained some interesting details regarding the species and their movements. Enough, however—thanks chiefly to the lighthouse-keepers—has been ascertained to prove that Ushant is a station of first-rate importance as an observatory for witnessing the movements of migratory birds; and thus the primary object of our visit was accomplished.

The Consul strongly advised us to quit France, lest we should experience further annoyance elsewhere, if we remained in Brittany, and we decided to proceed to the Channel Islands, selecting Alderney as being the most favourably situated for our investigations.

It should be remarked that Alderney is singularly destitute of trees, and therefore many of the resident and summer birds, which are natives of other islands of the group, are absent, while others are uncommon. Some of these, however, occur as birds of passage. Here, between 22nd and 29th September, we witnessed two decided movements—namely, on the 25th and 26th; when, among other species, the Mistle-Thrush, Ring-Ouzel, Goldcrest, Chiffchaff, Willow-Wren, Spotted Flycatcher, Pied Flycatcher, Turtle-Dove, Water-Rail, and Common Snipe appeared as immigrants. Unfortunately we were unable to say whether these birds arrived from the east after traversing the Channel coast of France, or from the north after leaving our own southern shores.

Some of our notes relating to the birds observed on the island are at variance with the experience of the late Mr Cecil Smith, the author of *The Birds of Guernsey and Neighbouring Islands of Alderney, etc.*, published in 1879. Such critical remarks as I have to offer will be found under the respective species in the subjoined list.

The following is a short account of all the species which came under notice at Ushant, and of some of those observed at Alderney :—

CORVUS CORAX, *Raven.*

Ushant.—The cliffs of Ushant afford a home and nesting-place for a pair of Ravens, whose presence was well known to the natives. These birds were seen daily in company ; but their young had, no doubt, been long since banished from the isle of their birth.

Sark.—Three Ravens were seen and heard under the east cliffs on 29th September. Mr Smith, writing in 1879, regards this bird as an occasional straggler, not a breeding bird on any of the Channel Islands. We had little doubt that the birds observed were natives.

STURNUS VULGARIS, *Starling.*

Ushant.—The lighthouse-keepers report that this is an extremely abundant bird of passage, and that it is also common as a winter visitor, especially during severe weather.

At the commencement of the third week of September, the Starling had not yet arrived at Ushant ; but we picked up a pair of withered wings of an unfortunate of the previous season.

PASSER DOMESTICUS, *House-Sparrow.*

Ushant.—A very common resident.

FRINGILLA CŒLEBS, *Chaffinch.*

Alderney.—Appears to be a decidedly uncommon species on the island: a fact which is probably due to the great scarcity of trees.

ACANTHIS CANNABINA, *Linnet.*

Ushant.—To the Linnet, Ushant offers considerable attractions in its numerous and enclosed furze-coverts, and the bird is consequently common. It was particularly abundant during our stay, when the ranks of the home birds were probably recruited by immigrants.

EMBERIZA MILIARIA, *Corn-Bunting.*

Ushant.—Is common and resident on the island.

EMBERIZA CITRINELLA, *Yellow Bunting.*

Ushant.—The Yellow Bunting is also a common and resident species.

ALAUDA ARVENSIS, *Skylark.*

Ushant.—Though most abundant during the autumn passage, the Skylark is also fairly common as a resident species. Great numbers are said to visit the island during severe winters, when the mainland is under snow.

MOTACILLA ALBA, *White Wagtail.*

Ushant.—Was numerous as a bird of passage during our sojourn on the island. It was to be seen in parties of from 20 to 30; and on some occasions we saw as many as 200 during our rambles. There were probably several hundreds on the island on certain days during our visit.

MOTACILLA LUGUBRIS, *Pied Wagtail.*

Alderney.—Was quite common, and observed, either singly or in family parties, between 22nd and 28th September. Mr Cecil Smith does not mention this species for Alderney, but remarks on its scarcity in Guernsey.

MOTACILLA BOARULA, *Grey Wagtail.*

Ushant.—Was far from common on the island; but several were observed on passage on 10th and 11th September, and carefully identified.

Alderney.—Not uncommon. This Wagtail is not mentioned for Alderney by Mr Cecil Smith.

MOTACILLA RAII, *Yellow Wagtail.*

Ushant.—Very common on passage during the time we were on the island—9th to 16th September. The birds observed were chiefly young of the year, but many fine old males were conspicuous among the bands of these migrants.

Alderney.—Very numerous as a bird of passage during our stay—22nd to 28th September. Mr Cecil Smith describes this bird as being only an occasional visitant on migration. This may be true for the other islands, but it certainly is not the case in Alderney; for which island, by the way, he does not mention this species.

ANTHUS PRATENSIS, *Meadow-Pipit.*

Ushant.—A very common bird on the island. It is a native species, but is most abundant as a bird of passage.

Alderney.—Very common during our stay.

ANTHUS OBSCURUS, *Rock-Pipit.*

Ushant.—Very numerous, and is, perhaps, a resident species, for the island affords most congenial haunts.

PARUS CÆRULEUS, *Blue Titmouse.*

Alderney.—Only once observed—namely, a single bird on 27th September. Mr Cecil Smith is quite right in describing this bird as by no means numerous in the Channel Islands.

REGULUS REGULUS, *Goldcrest.*

Alderney.—There were arrivals of this species on the nights of 25th and 26th September, and many were seen in the hedgerows on the following days. Mr Cecil Smith doubts whether the numbers of this bird in the Channel Islands are regularly increased in the autumn by migrants. There are, however, no suitable haunts for this bird in Alderney, and it was not seen prior to the immigration, which occurred on the night of 25th September.

PHYLLOSCOPUS COLLYBITA, *Chiffchaff.*

Alderney.—Several were observed and heard among the other immigrants which put in an appearance on 25th and 26th September. Mr Cecil Smith has no information regarding this bird in Alderney, where it is only a bird of passage.

PHYLLOSCOPUS TROCHILUS, *Willow-Warbler.*

Alderney.—Many observed on 25th, 26th, and 27th September, having arrived with other immigrants on the nights preceding the two first-named dates. This bird is not mentioned by Mr Cecil Smith for Alderney.

SYLVIA SYLVIA, *Common Whitethroat.*

Ushant.—Several whitethroats were seen; and the species is probably a summer visitor, as well as a bird of passage, on the island.

Alderney.—Was still common up to 28th September.

TURDUS VISCIVORUS, *Mistle-Thrush.*

Alderney.—On passage, were seen in the hedgerows in company with a number of immigrant species on 25th and 26th September.

TURDUS MUSICUS, *Song-Thrush.*

Alderney.—Many were noted on 25th and 26th September, having arrived on the island during the previous nights.

TURDUS ILIACUS, *Redwing.*

Ushant.—Four were seen, and their notes heard, on 10th September. This date is decidedly an early one on which to find this bird so far to the south. Many Redwings, however, depart from Iceland during the early days of this month.

TURDUS MERULA, *Blackbird.*

Alderney.—On 25th and 26th September, many immigrant Blackbirds, chiefly birds of the year, were observed, having arrived during the previous night.

TURDUS TORQUATUS, *Ring-Ouzel.*

Alderney. A single bird was seen on 22nd September. On the 26th several were observed in the hedgerows along with other immigrants, having arrived overnight. These birds were again noted on the 27th.

ERITHACUS RUBECULA, *Redbreast.*

Ushant.—The Redbreast is fairly common and resident. It was chiefly observed in the neighbourhood of houses, but was by no means confined thereto.

SAXICOLA ŒNANTHE, *Wheatear.*

Ushant.—Was very abundant as a bird of passage during our sojourn on the island—9th to 17th September. A few males showed traces of the summer dress, but the great majority of the birds seen were in russet plumage.

Alderney.—The 22nd to 28th of September, still common.

Pratincola rubetra, *Whinchat.*

Ushant.—A young male appeared on 11th September, and was the only bird of this species that came under our notice on the island.

Alderney.—Single birds were observed on 22nd and 27th September.

Pratincola rubicola, *Stonechat.*

Ushant.—This bird was surprisingly abundant; indeed, we have never seen this species in anything like the same numbers elsewhere. There must have been at least a couple of hundred of them on this small island. It is doubtless a resident species, and finds congenial haunts in the numerous compounds in which whin is cultivated.

Alderney.—Very common, 22nd to 28th September.

Accentor modularis, *Hedge-Accentor.*

Ushant.—Is a common resident, frequenting the whin-enclosures and the neighbourhood of the houses.

Troglodytes troglodytes, *Wren.*

Ushant.—The Wren was a common bird among the whin-enclosures, and is probably a resident on the island.

Muscicapa grisola, *Spotted Flycatcher.*

Alderney.—Was numerous in the hedgerows on 25th and 26th September, along with various immigrants, of which it was undoubtedly one. Mr Cecil Smith says that this bird "probably occurs in Alderney."

Muscicapa atricapilla, *Pied Flycatcher.*

Alderney.—One was distinctly seen by Mr Laidlaw and myself, independently, on the morning of 27th September. It was in a tall hedge, with Ring-Ouzels, Blackbirds, Goldcrests, Willow-Wrens, and other migrants, and was in "female" plumage. This species has no place in Mr Cecil Smith's book.

Hirundo rustica, *Swallow.*

Ushant.—Is a summer visitor, in limited numbers, to the island, and had not departed at the date of our exit.

Cotile riparia, *Sand-Martin.*

Alderney.—Not uncommon during the last week of September. Mr Cecil Smith regards this species as a spring visitor, not remaining to breed.

Alcedo ispida, *Kingfisher.*

Ushant.—This bird was surprisingly abundant all round the rocky coasts of the island, and appeared to be quite at home at the base of the highest cliffs, or on the margins of the surf-washed creeks. No doubt the extraordinary abundance of small fish and the crystal clearness of the sea were the attractions which had induced more than *fifty* of these birds to seek these singular island haunts. The great majority of the birds noted were adults, and were chiefly observed in pairs. These facts might lead one to consider that the Kingfisher was a resident species, breeding in more or less suitable places in the cliffs. This surely cannot be the case? It seems more probable that the bird is an autumn and winter visitor to Ushant. It was also numerous on the coast at Le Conquet.

Alderney.—Not uncommon on the coast.

Circus cyaneus, *Hen-Harrier.*

Ushant.—From 9th to 14th September, we saw a "Ring-tailed" Hen-Harrier daily. It chiefly frequented the cliffs, from which it occasionally made foraging excursions inland.

ACCIPITER NISUS, *Sparrow-Hawk.*

Ushant.—Single migratory birds were observed on several occasions between 10th and 16th September.

FALCO PEREGRINUS, *Peregrine Falcon.*

Ushant.—A pair is resident on the cliffs of the island, and the birds were observed almost daily.

Alderney.—On 20th September one was noted on the south cliff.

FALCO ÆSALON, *Merlin.*

Alderney.—One was seen at the west end of the island on 23rd September, doubtless a migrant.

FALCO TINNUNCULUS, *Kestrel.*

Ushant.—Migrants were not uncommon, and several were seen daily during our stay on the island—9th to 17th September.

Alderney. — Extremely abundant during the last weeks in September. Eight or nine were seen on the wing simultaneously, and a considerable number—a score or two—must have been present on several days. These birds were undoubtedly migrants, and were probably attracted by the abundance of a coleopteron which rejoices in the name of the "bloody-nosed beetle" (*Timarcha* sp.). The late Mr Smith doubted whether the numbers of this bird were at all increased during the migratory season in the Channel Islands. His remark certainly does not apply to Alderney, where we saw a score of different individuals in a few minutes on several occasions.

PANDION HALIAËTUS, *Osprey.*

Ushant.—On the 9th September an Osprey was observed quite close to our steamer, on leaving the island of Molène for Ushant.

PHALACROCORAX CARBO, *Cormorant.*

Ushant.—Was very abundant at the period of our visit. It is probably a winter visitor to Ushant and the adjacent islands, arriving early in the autumn.

Alderney. — Very abundant. Mr Cecil Smith regarded this species as uncommon in the Channel Islands, being replaced there by the Shag, a species which did not come under our notice on Alderney.

PHALACROCORAX GRACULUS, *Shag.*

Ushant.—Though the Shag was a common bird on the rocks around Ushant, yet it was not nearly so numerous as its congener the Cormorant. Like the latter bird, it is doubtless an autumn and winter visitor to the archipelago.

ARDEA CINEREA, *Heron.*

Ushant.—Single examples of the Heron were observed on 12th and 13th of September. This bird is probably only an occasional visitor, since there are few suitable shallows in which to capture prey.

COLUMBA LIVIA, *Rock-Dove.*

Alderney. — One was seen in company with some Jackdaws on the south cliffs on 24th September. Mr Cecil Smith never observed the Rock-Dove on any of the islands, though he was of opinion that a few might yet remain in Alderney.

TURTUR COMMUNIS, *Turtle-Dove.*

Ushant. — Single Turtle - Doves were observed on passage on 10th and 13th September.

Alderney. — A pair noted on 22nd, and a single bird on 25th September, were undoubtedly birds of passage.

RALLUS AQUATICUS, *Water-Rail.*

Alderney.—I twice put one out of a ditch on 26th September. There is an absence of suitable haunts for this bird on the island, and there can be little doubt that it was a recent arrival. Mr Cecil Smith makes no mention of this species for Alderney.

CHARADRIUS PLUVIALIS, *Golden Plover.*

Alderney. — Two were seen on 23rd September, several on 26th, and five were resting on the sands at Longey Bay on 27th. It is a bird of passage, and winter visitor to the island.

EUDROMIAS MORINELLUS, *Dotterel.*

Ushant.—Single examples were observed on 14th and 15th September; possibly the same bird was seen on each occasion. When first noted it was on the wing; on the second it allowed a close approach while resting, and flew off, uttering its unmusical note.

ÆGIALITIS HIATICOLA, *Ringed Plover.*

Ushant. — Was quite a common and doubtless a resident species on the island, finding on the extensive arid tracts a suitable breeding-ground, in the almost entire absence of littoral haunts.

VANELLUS VANELLUS, *Lapwing.*

Ushant.—Five arrived on the night of 9th September and remained during our visit. This species is mentioned by the light-keepers as being very common, both as a bird of passage and as a winter visitor to the island.

STREPSILAS INTERPRES, *Turnstone.*

Ushant. — This bird finds many congenial haunts along the rock-bound shores, and was quite common during our stay. It was often seen feeding inland, along with Ringed Plovers and a few Sanderlings, being attracted by the coleoptera which abounded on the island.

Alderney.—Fairly common.

HÆMATOPUS OSTRALEGUS, *Oyster-catcher.*

Ushant.—Common as an autumn migrant and winter visitor. It possibly also breeds, along with the Ringed Plover, on the arid, stony tract adjoining the shore on the west side of the island.

GALLINAGO CŒLESTIS, *Common Snipe.*

Alderney.—A single snipe was observed on 25th September. This bird is a winter visitor to the islands, but Alderney offers few, if any, attractions to this and other paludal species.

TRINGA ALPINA, *Dunlin.*

Alderney.—Two were seen on the sands of Longey Bay on 27th September.

CALIDRIS ARENARIA, *Sanderling.*

Ushant.—A few were observed during our stay. These were in small parties of from two to four individuals, and consorted with the Ringed Plovers and Turnstones, along with them seeking their food on the barren, stony land immediately adjoining the shore, on which a small species of beetle was numerous.

TOTANUS HYPOLEUCUS, *Common Sandpiper.*

Ushant.—Was extremely abundant as a bird of passage throughout our visit. It chiefly frequented the edge of the water, and was as numerous at the foot of the highest cliffs as elsewhere. On 10th September we computed that no fewer than 60 of these birds came under our notice. One was seen at sea between Le Conquet and Molène on 9th September.

Alderney.—A few seen daily on the coast up to 28th September, the day of our departure.

Totanus calidris, *Redshank.*

Ushant.—The island offered few inducements to the Redshank, and hence one or two, seen singly, were all that we observed during our sojourn.

Numenius phæopus, *Whimbrel.*

Ushant.—This bird was seen daily on passage on all the coasts of the island, and was quite an abundant species.

Numenius arquata, *Curlew.*

Ushant.—Was one of the most numerously represented species during our visit. It was to be seen daily in parties, sometimes 100 strong, feeding on the parched land, where small beetles were very numerous at the time, whose presence, no doubt, induced these birds to tarry on the island in such numbers.

Sterna macrura, *Arctic Tern.*

Ushant.—Terns were abundant off the island during our visit; but it was not until 14th September that we identified *Sterna macrura.* On that day we watched a number of Arctic terns as they rested on a rock just off the south coast of the island. Among these we detected no fewer than ten individuals in that stage of adolescent plumage which led Mr Ridgway to describe this bird as a distinct species under the name of *Sterna portlandica.* There was no doubt as to their identity, for, as if to oblige us, some of them continually left the rock and alighted on the shore just at our feet. We may, I think, fairly assume from this, that there were many more of these interesting birds around Ushant and the neighbouring islands.

STERNA FLUVIATILIS, *Common Tern.*

Ushant.—Very common. Many old birds were still observed feeding their young, while the latter rested on the rocks off the coast—a fact which indicates that there is a breeding-station near at hand, perhaps on the Isle of Balanec, north of Molène, where, we were informed, the "Hirondelle de mer" bred in some numbers.

STERNA MINUTA, *Little Tern.*

Ushant. — Was observed off Molène on 9th September, and was not uncommon off Ushant during our visit.

STERNA CANTIACA, *Sandwich Tern.*

Ushant.—Not uncommon throughout our stay, and also observed off Molène.

LARUS CANUS, *Common Gull.*

Ushant.—Not uncommon on the coast and on the outlying rocks, to which it is an autumn and winter visitor.

LARUS ARGENTATUS, *Herring Gull.*

Ushant.—Common off the coast and also at Molène, where it is an autumn and winter visitor.

LARUS FUSCUS, *Lesser Black-backed Gull.*

Ushant.—This was quite common on the coast, and was also observed at Molène.

LARUS MARINUS, *Great Black-backed Gull.*

Ushant. — Fairly common off the island and at Molène.

RISSA TRIDACTYLA, *Kittiwake*.

Ushant.—Young and old Kittiwakes were not uncommon on the stacks and rocks after 14th September, on which day they made their appearance off the island, along with the Arctic Tern.

Alderney.—A number seen off the island on 28th September.

PUFFINUS ANGLORUM, *Manx Shearwater*.

Ushant.—This Shearwater was very abundant on 9th and 17th September, when we were en route for, and returning from, Ushant. It was chiefly observed when some little distance from land.

Alderney.—This bird was very numerous off the Casquets on 30th September. Mr Cecil Smith tells us that it is an occasional wanderer to the Channel Islands.

PUFFINUS GRAVIS, *Great Shearwater*.

Ushant.—When just off Ushant on 17th September we saw six Great Shearwaters, either singly or in pairs, fairly close to the steamer.

Alderney.—Off the Casquets, on 30th September, one was seen along with the Manx Shearwaters.

INDEX

G

K

L

M

N

O

Q

R

S

T

PRINTED BY
OLIVER AND BOYD
EDINBURGH

Printed in the United States
By Bookmasters